W9-BIP-644

RETURN TO DIVERSITY

RETURN TO DIVERSITY

A Political History of East Central Europe Since World War II

Joseph Rothschild

New York Oxford
OXFORD UNIVERSITY PRESS
1989

Oxford University Press

Oxford New York Toronto
Delhi Bombay Calcutta Madras Karachi
Petaling Jaya Singapore Hong Kong Tokyo
Nairobi Dar es Salaam Cape Town
Melbourne Auckland

and associated companies in
Berlin Ibadan

Published by Oxford University Press, Inc.,
200 Madison Avenue, New York, New York 10016

Library of Congress Cataloging-in-Publication Data
Rothschild, Joseph.
Return to diversity : a political history of East Central Europe since World War II / Joseph Rothschild.
p. cm.
Bibliography: p.
Includes index.
ISBN 0-19-504574-2
[1. Europe, Eastern—History—1945-] I. Title.
DJK50.R67 1989
949—dc19 88-1505
CIP

2 4 6 8 9 7 5 3 1

Printed in the United States of America
on acid-free paper

The struggle of man against power
is the struggle of memory
against forgetting.

MILAN KUNDERA

This book is dedicated to Sir William Deakin and to the memories of Henry L. Roberts and Hugh Seton-Watson, teachers who introduced me to the study of East Central Europe.

Preface

The main challenge in writing a basic general survey such as I attempt here is distillation. And distillation inevitably entails the omission or condensation of material that specialists would prefer to include or to develop. But a basic book cannot be a comprehensive chronology, nor a heavily footnoted research monograph, nor a deeply searching analysis. While I, too, regret the absence from the following pages of many interesting episodes, important personalities, and suggestive arguments that I would have liked to include, I must nevertheless ask the reader to judge this volume by the criteria that are appropriate to its own genre—that of a general survey—and not by standards alien to it.

A word is in order as to why this book does not include a consolidated analysis of the political history of East Germany comparable to my extended probing into those of Poland, Czechoslovakia, Hungary, Yugoslavia, Romania, Bulgaria, and Albania. The (East) German Democratic Republic is indeed a state, but it is not a nation and is less than half a country. Before World War II it was not even a state (unlike my other cases), but simply a part of Germany and hence not in East Central Europe. Since I view this book as a continuation of my *East Central Europe Between the Two World Wars* (1974), those are valid reasons for omitting East Germany here. More important, however, is the consideration that East German domestic and foreign politics are so overwhelmingly a part and a function of "the question of divided Germany" on the Great Power agenda, that any serious effort on my part to explore them in this book would have burst its perimeters and muddied my professional waters. Greece, in turn, is omitted because it is essentially a Mediterranean, not an East Central European, country in its cultural, economic, and political perspectives and because it was not subsumed into the Communist orbit after World War II. Analogous reasons account for the omission of Austria. Finally, the three Baltic republics of Estonia, Latvia, and Lithuania are left out because their brief interwar period of formal independence did not survive World War II.

It is conventional for the author of a professional book to list a number of colleagues who have helpfully read all or parts of the prepublication manuscript, to thank them, and to avow a pious acknowledgment that any residual flaws and errors are nevertheless the author's own responsibility and are not to be imputed to these readers. Deeming this convention to be saccharine and somewhat hypocritical, I do not abide by it. My expressions of public, and very sincere, gratitude are here reserved for my supportive family; my splendid typist, Audrey McInerney; and the helpful officers of two valued institutions who invited, encouraged, and generously subsidized my work: Drs. Enid C. B. Schoettle and Paul Balaran of the International Division of The Ford Foundation and Dr. Jason H. Parker of the American Council of Learned Societies.

New York J. R.
January 1988

Contents

RETURN TO DIVERSITY

EAST CENTRAL EUROPE
Miles
0 100 200 300
BARENTS SEA
Murmansk
WHITE SEA
Archangel
FINLAND
NORWAY
Oslo
SWEDEN
Helsinki
Stockholm
Leningrad
NORTH SEA
Tallin
ESTONIAN S.S.R.
BALTIC SEA
DENMARK
Riga
Copenhagen
LATVIAN S.S.R.
LITHUANIAN S.S.R.
Vilnius
Minsk
Gdansk
Kaliningrad
EAST GERMANY
Berlin
Bonn
POLAND
Warsaw
BYLORUSSIAN S.S.R.
WEST GERMANY
Prague
CZECHOSLOVAKIA
Danube
Munich
Vienna
AUSTRIA
Budapest
HUNGARY
Milan
Trieste
Kiev
Dnieper
UKRAINIAN S.S.R.
MOLDAVIAN S.S.R.
Kishinev
Odessa
ROMANIA
Sevastopol
YUGOSLAVIA
Belgrade
Bucharest
Danube
ITALY
ADRIATIC SEA
Rome
Prishtina
BULGARIA
Sofia
Tirana
ALBANIA
GREECE
Istanbul
TYRRHENIAN SEA
BLACK SEA
TURKEY
Ankara
UNION OF SOVIET SOCIALIST REPUBLICS
Moscow

1

The Interwar Background

1

At the close of World War I, the four defeated empires that had dominated and ruled East Central Europe—the German, Habsburg, Ottoman, and Russian empires—were replaced by a dozen new or restored or enlarged would-be nation-states, all of which based their asserted legitimation on the then reigning politico-moral principle of national self-determination. Though the territorial arrangements of 1919 to 1921 still left a number of additional nations in East Central Europe stateless and created problems of aggrieved minorities allocated to states toward which they felt little or no affinity (conditions that induced revisionist apologists for the territorial losers of World War I to charge that the territorial arrangements were merely a cynical and unprincipled victors' fiat), for all their admitted flaws, they still freed three times as many people from nationally alien rule as they subjected to such rule. The real political weakness of the interwar effort to implement the principle of national self-determination in East Central Europe lay not in its alleged hypocrisy, but in the impossibility of reconciling it with three other important aims of the peacemakers of 1919 to 1921: the permanent diminution of German power, the permanent containment of Russian power, and the permanent restoration of international order in Europe. In other words, the geopolitical map of interwar East Central Europe, with its plethora of new, restored, and enlarged *soi-disant* nation-states, was not congruent with the real distribution of power in Europe.

Germany and Soviet Russia embodied the two basic revisionist threats to the territorial and social settlements of the interwar years. Though most East Central European regimes of the time were more mesmerized by the Bolshevik threat, Germany proved to be the more active menace and for that reason we focus on it first.

The defeat of Germany in 1918 was deceptive. Neither in absolute nor in relative terms had Germany been weakened to any thing like the extent that was often assumed in the 1920s. In absolute terms, Germany's industrial and transportation resources had been left largely intact because World War I had not been fought on its territory. In relative terms, a territorial settlement predicated on the national principle, such as that popular in 1919 to 1921, ipso facto left Germany as Europe's second largest country after Russia. Relative to East Central Europe, Germany gained through the replacement of the Habsburg Empire, which for all its infirmities had still been a major power, as a neighbor by a large number of frail and mutually hostile successor states in the Danubian area to the southeast, and through the substitution of Poland and the Baltic States for Russia as its immediate eastern neighbors. Germany's geographical position in the center of the Continent was only enhanced by these developments. The very existence of the newly independent but highly vulnerable states of East Central Europe, endorsed by the victorious Western Allies, proved on balance a political and diplomatic asset to Germany. It (1) initially buffered Germany against a spillover of the Bolshevik Revolution, (2) then tempted Soviet Russia to collaborate with Germany throughout the 1920s and in the partition of this area in 1939 and 1940, and (3) ultimately frustrated efforts at Soviet–Western cooperation to halt Nazi Germany in the late 1930s, as the West was then inhibited by its commitments to the successor states from paying the Soviet Union's price for such cooperation—the sacrifice of East Central Europe's effective independence to Soviet hegemony.

The governments of Weimar Germany pursued a "Prussian" policy of directing the brunt of their revisionist pressure against interwar Poland, in the hope of recovering at least a substantial part, if not all, of the prewar Reich frontiers there. Hitler, on the contrary, contemptuously dismissed as inadequate such a limited program. Setting his sights on the conquest of all East Central and Eastern Europe, he temporarily froze the German–Polish revisionist issue with the bilateral Non-Aggression Statement of January 26, 1934, and launched his program of virtually limitless conquest by first following the "Austrian" pattern of establishing hegemony over the Danube Valley. Austria and

Czechoslovakia, rather than Poland, thus became his initial international victims.

It has often, and correctly, been pointed out that the Nazi concept of race was politically incompatible with the existence of independent East Central and Eastern European states. Less attention has been given to the at least equally sinister concept of space in Hitler's politico-ideological armory. While racial rhetoric was occasionally used by certain Nazis (other than Hitler) to flatter the supposedly "young" and "vigorous" peoples of East Central Europe into deserting their allegedly "decadent" and "enfeebled" Western allies and patrons, the political language of space always implied conquest and reduction to peonage of the peoples to Germany's east and southeast. Indeed, the capacity for such spatial expansion was defined as the test and measure of racial vitality.

Given his maximalist program of expansion and conquest, Hitler was tactically correct in identifying Czechoslovakia, rather than Poland, as the keystone of Germany's "encirclement" that would have to be dislodged first to collapse that arch. Territorial revisionism against Poland was likely to be more limited in its political effect, since it would initially have to be coordinated with Soviet Russia; it implied shared influence rather than exclusive domination. Against Czechoslovakia, Hitler's ally would be a Hungary conveniently revisionist but too weak to present a serious obstacle to further German expansion. Furthermore, the German officer corps, still heavily "Prussian" in its political outlook and self-image, might be satisfied with the defeat of Poland and thereafter reluctant to be used for further Danubian, Balkan, and Russian conquests, toward which it was historically conditioned to be either indifferent or even unfriendly. Finally, Czechoslovakia, unlike Poland, could be conveniently tarred with the phony but propagandistically effective brush of serving as "Bolshevism's Central European aircraft carrier" by virtue of the Czechoslovak-Soviet Pact of May 16, 1935, which supplemented the Franco-Soviet Mutual Assistance Treaty of May 2, 1935. Though this pair of agreements had been made in response to Hitler's reintroduction of German conscription on March 16, 1935, in violation of the Versailles Treaty, and though they were soon to be tested and found wanting by Hitler's remilitarization of the Rhineland on March 7, 1936—again in violation of treaty obligations—which rendered France's military commitments to its several East Central European allies strategically worthless, the German propaganda assault on Czechoslovakia proved successful. Its victim stood isolated, friendless, and shunned by all its neighbors at the time of the Munich tragedy in September 1938.

East Central European anti-Communism and fear of Soviet ambitions thus benefited and were manipulated by Germany—to such an extent, indeed, that the international politics of the 1930s were fatally skewed by fundamental misjudgments about the source of the immediate threat to the area's independence. A number of the local states owed all or much of their territory to Russia's weakness between 1917 and 1921; the ruling elites in all of them feared Communism. Hence they were understandably reluctant on the eve of World War II to grant the Soviet army access to their countries as their contribution to collective security against Nazi Germany. They feared that once in, the Soviets were unlikely to depart, least of all from lands that had been part of the Russian Empire. The Western governments, in turn, sharing many of these ideological and political anxieties and committed to the principle of the integrity of small states, were reluctant to press them into such a hazardous concession. Stalin, however, could scarcely be impressed by the West's assertion against the Soviet Union in mid-1939 of a principle—the territorial inviolability of sovereign states—that it had indecently sacrificed to Hitler at Munich less than a year before.

A circular dilemma thus arose: the East Central European governments were unwilling to accept Soviet assistance against the Nazi threat lest it either provoke the German invasion that collective security was intended to deter or simply become a Soviet occupation; the West now refused to cap its abandonment of Czechoslovakia in 1938 by coercing Poland and Romania into abdicating their sovereignty to the Soviet Union in 1939; Stalin was unwilling to expose his country to the risk of bearing the brunt of a war against Germany unless he could at least reduce that risk by forestalling Hitler in a military occupation of East Central Europe. Underlying the failure to resolve this dilemma were a set of interlocking misjudgments: Stalin was skeptical of the West's readiness finally to stand up to Hitler, underestimated Britain's military competence, and overestimated France's military prowess. The Western governments depreciated the Soviet Union's military value and presumed that ideological incompatibility would prevent any Nazi–Soviet rapprochement. All miscalculated. The upshot of the unresolved dilemma was the German-Soviet Pact of August 23, 1939, and World War II, in which the *Wehrmacht* quickly disposed of the Polish and French armies and thus destroyed that Continental second front for which Stalin was to implore his allies when that same *Wehrmacht* was later turned against him. A moral of this sad tale is that the balance of power is never automatic but requires rationality, perceptiveness, judgment, and perhaps even wisdom for its proper recognition.

The ease with which Germany, and later Russia, regained control over interwar East Central Europe was based on more than just ideological and psychological manipulation, important though that was. They also capitalized on the abdication of the other Great Powers and on the profound politico-demographic and socioeconomic weaknesses and conflicts within the area itself. On the morrow of the peace settlements closing World War I, the United States withdrew into isolation, the United Kingdom turned to a policy of encouraging the revival of Germany in order to "correct" a supposed, but actually illusory, French Continental preponderance, Italy entertained its own dreams of hegemony in the Balkan Peninsula and the Danube Valley, and France adopted a self-contradictory stance of making far-ranging political and military commitments to several states in East Central Europe but simultaneously undermining these commitments with defensive and isolationist strategic and economic postures. France, though granting them some loans, traded very little with its East Central European protégés, protected its own agriculture from their surpluses, and sought to veto their industrialization programs for refining their own mineral resources owned by French concessionaires. Simultaneously, France's Maginot strategy—a function of the multiple trauma of having been bled white during the war and then deserted by one ally (the United States) and persistently restrained by the other (the United Kingdom) after its close—eroded the credibility of its alliance commitments in East Central Europe. That credibility was finally flushed away with its passive acceptance of Hitler's remilitarization of the Rhineland, after which he could direct the bulk of the *Wehrmacht* against selected East Central European victims without fear of French counteraction in western Germany.

Thus East Central European hopes of achieving security by bringing the weight of benevolent, if distant, Great Powers to bear against the area's rapacious and immediate neighbors proved abortive. During the 1920s, only Germany's and Russia's temporary postwar and postrevolutionary exhaustion had provided East Central Europe with a respite, despite their ominous diplomatic collaboration. In the 1930s, though both countries were rapidly reviving, their ideological and political enmity briefly extended this reprieve to the lands between them, until their fateful reconciliation at the area's expense in 1939.

2

Given this constellation of predatory, indifferent, and ineffective Great Powers, a constellation that it could neither prevent nor even control, East Central Europe might have attained at least a minimum power credibility if it had been able to achieve internal regional solidarity and some system of mutual assistance. But this alternative, too, was negated by the multiple divisions and rivalries that were born of competing territorial claims, ethnic-minority tensions, socioeconomic poverty, mutually irritating national psychologies, and sheer political myopia. These factors transformed the area's internal relations into a cockpit and facilitated Hitler's program of conquest. It is scarcely an exaggeration to suggest that as a general rule in interwar East Central Europe, common borders entailed hostile relations. Thus the "blame" for the demise of the region's independence must be charged to its own fundamental weaknesses, the instability of its institutions, and its irresponsible governments, as well as to the active and passive faults of the Great Powers.

Simply to list the area's internal irredentist disputes may convey an impression of their cumulative complexity, though not of their bitter and well-nigh paralyzing intensity. Lithuania and Poland quarreled over Wilno (Vilnius, Vilna), which the former claimed on historical and the latter on ethnodemographic and strategic grounds. Poland and Czechoslovakia were mutually alienated by (1) their dispute over Teschen (Těšín, Cieszyn), where the former's sounder ethnodemographic claims clashed with the latter's economic needs; (2) their contrasting perceptions of Russia's and Hungary's proper roles in the European balance, each regarding the other's *bête noire* with some benevolence; (3) the conviction of each that the other had doomed itself by greedily incorporating too many unabsorbable, and hence inflammable, ethnic minorities; and (4) their contrasting social structures and national psychologies—that is, Polish gentry versus Czech bourgeois. Czechoslovakia was also under revisionist pressure on historical and ethnodemographic grounds from Hungary. Hungary, as the biggest territorial loser of World War I, nursed territorial claims on historical and/or ethnodemographic grounds against all four of its neighbors: Czechoslovakia in regard to Slovakia and Ruthenia; Romania over Transylvania; Yugoslavia with reference to the Vojvodina and perhaps Croatia; Austria over the Burgenland (this last, less intensely than the others). Yugoslavia coveted the Slovene-populated portion of Austria's Carinthian province, and Yugoslavia and Romania were, in turn, the objects of Bulgarian irredentist resentments over Macedonia and Southern Dobruja, respectively. In addition, Bulgaria directed similar pressures against Greece over parts of Macedonia and Thrace. Bulgaria's revisionist rationale was the characteristic combination of historical, ethnodemographic, economic, and strategic arguments. As regards Albania and Austria, finally, the major problem was not that irredentist aspirations were harbored by and against them—though they, too, existed—but that their very existence was challenged and their survival seemed doubtful during the interwar era.

As though these quarrels within the region were not enough, a number of its states were under even more ominous pressures from the Great Powers. Weimar Germany remained unreconciled to the loss of the Pomeranian "Corridor" and of southeastern Silesia to Poland, and Hitler was to add to these revisionist grievances his claims to Czechoslovakia's highly strategic, German-populated, Sudeten perimeter and to all of Austria. Less pressing was Germany's suit against Lithuania for the retrocession of the city and district of Klaipėda (Memel). The Soviet Union remained openly unreconciled to Romania's incorporation of Bessarabia and harbored designs on Poland's eastern borderlands, with their large Belorussian and Ukrainian ethnic populations. Its attitude toward the Baltic States was more complex but still ambivalent. Italy craved Yugoslavia's Dalmatian littoral on the Adriatic Sea and schemed to fragment the entire Yugoslav state into its ethnoregional components. It also aspired to control Albania directly and to intimidate Greece into subservience. Indeed, Italy's ambitions also included the establishment of diplomatic protectorates over Austria, Hungary, and Bulgaria, in order to redouble the pressure on Yugoslavia. But in contrast to Germany and the Soviet Union, Italy lacked the economic and military muscle to realize its political designs.

Thus each state of interwar East Central Europe had one or more enemies within the area, and each of the "victor" states among them also had a Great Power enemy—Poland even had two. Its numerous "internal" enmities, alas, rendered the region even weaker than it need have been with respect to the "external" ones, and all efforts at reconciling the former were aborted by rampant chauvinism. The spirit of the age was not supranational, as had been naïvely predicted during the war, but ultranational. Indeed, it appears that the only really potent international ideology in the area at that time was neither Marxism, on the left hand, nor dynastic loyalism, on the right, but anti-Semitism based on both conviction and expedience. This, in turn, provided a bond and precondition for eventual collaboration with the Nazis, including the administration of wartime genocide.

3

An important, often the main, component of the several revisionist-irredentist territorial disputes in interwar East Central Europe was the ethnic one—specifically, one state's interest in politically "redeeming," or at least culturally sustaining, a minority of its own nationality that happened to live in another state, and the "host" state's indignant repudiation of what it regarded as illicit pressures on its territorial integrity or internal sovereignty. Admittedly, the existence of ethnic minorities was nothing new in East Central Europe. But before World War I, the region had by and large been ruled by empires. Ethnic minorities generally fare better in empires whose central elites seek to impose political order but are rather indifferent to the ethnosocial and ethnocultural heterogeneity of the subject populations than they do in would-be nation-states whose central elites view themselves as the custodians of explicit and specific national cultures that serve them as comprehensive integrationist ideologies and programs. And, indeed, the interwar states of East Central Europe, unlike the Ottoman Empire and the Austrian half of the Hapsburg Empire, regarded themselves as explicit and specific nation-states. Hence the lot of the numerous and vocal interwar ethnic minorities was often emotionally more demeaning and politically more hopeless than had formerly been the case. Thus, for example, the Czechs, Poles, Slovenes, and other ethnic groups of the Habsburg Empire had not been obliged to consider themselves as subordinate minorities in an explicitly German state. Though they had felt themselves ethnically aggrieved at various times, they could always quite realistically anticipate a future imperial government's reversal of the schedule of ethnic favoritism. Even the more consistently excluded ethnic minorities of the empire's Hungarian half awaited a change with the next royal succession. But in the so-called nation-states of the interwar era, an ethnic minority seemed fated, short of a war and a redrawing of frontiers, to remain disadvantaged forever, not simply in the neutral statistical sense, but also in terms of political, economic, cultural, and sometimes even civil and legal deprivations. Hence it tended to seek succor from its ethnic and cultural "mother country" against the pressures of the "host" state, and thus the dispute was internationalized. The Jews, being without a state of their own, lacked this option and so felt particularly exposed politically.

The "host" government, in turn, was committed to the promotion of the specific national culture of its state-nation throughout its territory.

Otherwise, it reasoned, the achievement of national independence would have been purposeless. Its apprehensions about "subversion" tended quickly to become as exaggerated, albeit sincere, as the ethnic minority's fears of "extinction." The resultant reciprocal recriminations would become particularly truculent, the protagonists' stances particularly rigid, and the quarrel particularly dangerous if, as was often the case, the ethnic minority and the interested "mother country" to which it appealed represented one of the region's prewar dominant powers—Germany, Hungary, Bulgaria (Macedonia), Russia (Ukraine)—still unreconciled to its recent defeat and loss.

The determination of a newly independent state to "nation-ize" not only its cultural and political patrimonies, but also its economic wealth was often a key motive behind such seemingly social and "class" programs as land reform and etatist industrialization. They were politically easiest where the property of "alien" landlords and entrepreneurs could be expropriated for the benefit of "native" peasants and bureaucrats. Such an amalgamation of ethnic and social policy was facilitated by the fact that ethnic, religious, and class differences and identifies often coincided or at least overlapped. Thus in Poland, the Baltic States, and the former Habsburg lands, the large estate owners were Poles, Germans, and Magyars, while the entrepreneurial class was heavily German and Jewish and only partly native. In the Balkans, the entrepreneurial class was Greek, Italian, and Jewish and only incipiently native, while in several areas, the landlords were still Muslim or Magyar. Another indirect way of implementing ethnic policy in the absence of explicit legislative authorization, which was generally avoided for legal reasons or with an eye to public relations, was through silent but relentless administrative discretion. All in all, the importance of ethnic consciousness in the new, restored, or enlarged victor states of interwar East Central Europe is illustrated by the observation that none of them experienced the sharp social and class violence that on the morrow of World War I wracked the losers—first Russia, then Germany, Austria, Hungary, and Bulgaria.

Standing politically midway between state-nations and subordinate ethnic minorities were those peoples who were officially defined as belonging to the former but felt themselves not only culturally distinct from, but also politically and economically exploited by the dominant part of that state-nation. The most vivid interwar examples of these groups were the Slovaks with respect to the Czechs, and the Croats with respect to the Serbs. Each aggrieved group became increasingly disenchanted with and suspicious of the formal ideology of "Czechoslovak" and "Yugoslav" nationality, which appeared to it to be a manipulative device screening, respectively, Czech and Serb domination. Whereas in Czechoslovakia there was a correspondence between the Czechs' political control and their superior economic and cultural resources vis-à-vis the Slovaks, in Yugoslavia there existed a "crossed" relationship between Serb political domination, on the one hand, and the more advanced and developed Croatian economic and cultural levels, on the other.

All in all, the rather complicated texture of the ethnic-minority question both reflected the attempted but fragile interwar European power balance and, due to the ensuing political tensions, helped to overturn it. These chronic tensions, and particularly the manner in which Nazi Germany manipulated them, elicited a sharply different approach to the entire problem at the close of World War II. Whereas at the end of the first world conflagration, there had been many frontier changes but relatively little mass population movement in East Central Europe, after the second one, there were fewer frontier changes—the major exceptions being in regard to the Soviet Union's western borders and Poland's eastern and western ones—but enormous population migrations and expulsions, following the Nazi genocide of the area's Jewish and Gypsy minorities and persecution of several indigenous nations. Hitler, having both rendered the numerous German minority in East Central Europe odious to the Slavic peoples and demonstrated the ease with which minorities could be eliminated, thereby provoked the colossal enforced *Völkerwanderung* of 1944 to 1946. In the course of this migration, the millennium-long German eastward expansion by peasant, burgher, miner, monk, and soldier was reversed, and the political achievements of Henry the Lion, Frederick the Great, and Bismarck were undone. While proclaiming that he only wished to save Europe from the supposedly corrosive "Internationals" (Communist, Jewish, Jesuit, Masonic, plutocratic, and so on), Hitler had in fact persuaded the 6 million *Volksdeutsche* of East Central Europe to serve him as a truly subversive Pan-German "International," to their ultimate misfortune.

4

While ethnonational tensions constituted interwar East Central Europe's most vivid and sensitive political problem and were, indeed, often exploited in order to obscure social and economic weaknesses, these weaknesses proved just as chronically debilitating and difficult to correct. By virtually every relevant statistical index, East Central Europe was less productive, less literate, and less healthy than West Central and Western Europe. A potentially rich region with poor people, its interwar censuses record not so much a distribution of wealth as a maldistribution of poverty. The main component of this sad spectacle was the so-called peasant question, in both its socioeconomic and its ideological manifestations.

Interwar East Central Europe was preponderantly agricultural. While a far higher proportion of its population was engaged in farming than was the case in Western Europe, the productivity of its agriculture in terms both of yield rates per unit of agricultural area and of yield rates per agricultural worker was far lower. The result was a vicious cycle of rural undercapitalization, underproductivity, underconsumption, underemployment, overpopulation, and pervasive misery. Despite strenuous, if often misapplied, efforts to correct these imbalances and to increase the area's wealth through industrialization, in 1938, on the eve of World War II, East Central Europe still produced only 8 percent of the industrial output of all Europe minus the Soviet Union, and of this small share, one-third was recorded by Czechoslovakia. Except in that country, whose western half was the area's most thoroughly industrialized region, the fate of the several states' economies was annually determined by the single, hazardous, factor of weather.

Problems ancillary to and aggravating this low productivity in the agricultural sector were weak transportation systems, disruption of prewar trade patterns, economic nationalism and competitive striving for autarky (especially prominent and destructive during the Great Depression of the early 1930s), competition of Argentine and North American grains in the markets of Western Europe, and drastic reduction of opportunities for emigration to the United States. The region's swelling surplus peasant population vegetated at bare subsistence levels on its holdings, subdividing them into ever smaller and less rational plots. The peasantry's very existence and condition of underemployment discouraged any investment in agronomic technology. Even then, the undernourished peasants were scarcely permitted to consume an adequate proportion of their relatively low food output because government fiscal, tariff, and investment policies consistently forced them to sell at a pittance far more than any authentic surplus of their produce in order to raise cash for the payment of taxes, debts, fees, and a few astronomically priced (because protected and cartelized) essential industrial products.

Where governments did arrange land reforms for the ostensible benefit of the peasantry, the motivation and hence the application were primarily political—either, as mentioned above, to expropriate the property of ethnically "alien" landlords or to immunize a restless peasantry against the feared attractions of Communism—and were not adequately supplemented with equivalent interest in correcting the economic and agronomic malaise of agriculture. The peasant's standard of living was falling precisely at a time when his expectations and self-esteem were rising. His travels and other experiences as a mobilized soldier in World War I not only had sophisticated his material wants, but also had shown him the extent to which governments and urban populations depended on his docility and labor. He now responded to his interwar lot by vacillating among resentment, mistrust, despair, and rage. In particular, the combination of his disastrous impoverishment during the Great Depression of the early 1930s, when the industrial–agricultural price scissors opened drastically against him, and his economic rescue—in the Danubian and Balkan countries, if not in Poland and Czechoslovakia—through Nazi Germany's bulk purchase at high prices of his produce in the late 1930s served to radicalize the East Central European peasant—occasionally toward the Left but more frequently toward the new Radical Right. This trend suggests that a reexamination of the traditional claims and postures of ideologues and politicians of the peasantry is in order.

Against these ideologues' claims, in the tradition of Rousseau and Jefferson, that the peasant's proximity to nature, rustic life style, and sustained work habits allegedly made him a "naturally" democratic, tolerant, peaceable, and cooperative citizen,[1] we may offer the suggestion that the Eastern European and East Central European peasant's characteristic political behavior, as expressed by long periods of submissiveness interspersed with bouts of *jacquerie* violence, indicates profound, albeit understandable, apathy, alienation, and rancor. Excluded from the general progress of Europe, he felt himself to be both the guardian and the victim of anachronistic values and institutions, whose very anachronism undermined and negated the potential power of the peasantry as the area's most numerous class. The peasant's political stance in the interwar era was problematic and uneasy. Grateful for land reform, he also resented that one of its side effects had been to intensify the control of the state apparatus over his village. This control he felt to be exploitative rather than benevolent, exercised in its own interest by a culturally alien urban bureaucracy that would either neglect or suppress but neither probe nor solve the social tensions accruing from the economic malaise of the countryside.

Against its ideologues' rhapsodic presentation of peasantism as a supposed humanistic alternative to allegedly crassly materialistic capitalism and socialism, we may legitimately note their naïveté about both the "soulless" industrialism espoused by these two competing ideologies and their own favored "peasant way of life." For the hard fact is that the peasant could achieve prosperity only by transforming that way of life into an integrated, productive relationship with urban market needs and industrial capacities. Furthermore, the ideological celebrators of the peasantry appear to have misread or misrepresented the real views of their claimed constituency. For the peasant's actual attitude toward industrialization was less one of hostility than one of ambivalence: he was both fascinated and afraid. He realized that it alone held out the promise of salvation from rural poverty and overpopulation. But he also dreaded industrialization as a threat to his values and traditions. More specifically, he shrewdly suspected that its immediate costs in terms of restricted consumption and increased prices and taxes would be unloaded onto his shoulders—or, rather, squeezed from his belly.

The general peasant resentment and mistrust of urban society extended to the proletariat, the area's other interwar "outsider" class. The East Central European Socialist parties and workers, in turn, feared and shunned the peasant masses as incarnating an allegedly reactionary, clericalist threat to economic and social progress. It was, indeed, true that the only political parties other than explicitly peasant ones that the East Central European peasantry occasionally supported were explicitly Christian-denominational ones in the 1920s and also Radical Right ones in the Depression decade of the 1930s. More generally, the area's still young and small urban proletariat, in its anxiety to avoid being weakened or manipulated from any quarter, tended to isolate itself from social alliances with any older and larger classes, even on the rare occasions when such groups were available as would-be allies.

The potential political power placed in the peasantry's hands by the universal suffrage introduced throughout interwar East Central Europe, except in Hungary, was soon blunted by the emergence of a specific political ruling class. This class initially coopted peasant political leaders and eroded the peasantist component of their political commitments. Later, alarmed by escalating social and political unrest, it simply replaced the formally democratic political institutions with authoritarian ones, except in Czechoslovakia. This political ruling class was not, contrary to conventional assumptions, the bourgeoisie, which was quite weak and either dependent on state subsidies or ethnically "alien" and hence vulnerable. Rather, it was the bureaucracy, which was allied with and recruited from the intelligentsia.

The conduct of peasant political leaders ran the gamut from "bearer of the national conscience" or "peasant Gracchus," through "statesman," "pragmatist," and "power broker," to sheer "betrayer of trust" and "office seeker" or, alternatively, "opposition demagogue." None adequately benefited their village constituencies, which were nevertheless pathetically loyal to them. In many cases, the peasant politician's class pride was accompanied and corroded by a residual political inferiority complex. This led him to overvalue the fact of his admission into the councils of government, where his often vague programs and generalized aspirations were promptly and easily neutralized by cabinet colleagues who appealed to his sense of "realism" or "patriotism." There were always plausible reasons, for example, why indirect taxes on necessities consumed in the village were more "feasible" than direct taxes on the incomes of the urban entrepreneurial, professional, and bureaucratic classes, or why the "national interest" required that the resultant revenue he spent on the army and on subsidized industry rather than reinvested in agriculture or in rural amenities. While prominent peasant "tribunes" were often thus coopted at the top, their party machines were always infiltrated at the less visible middle echelons by the political class of lawyers and bureaucrats that had already captured control of other political parties and of the state apparatus as such.

International peasant solidarity was articulated in the interwar era by the so-called Green International, an appellation intended to symbolize its supposed historical role as an alternative to the "Red" International of Communists and the "White" International of capitalists and landlords. Its institutional expression was the International Agrarian Bureau, established in Prague by several East Central European peasant parties. Organizationally and financially, it was controlled by the Czechoslovak Agrarians, who sought to give it a Slavophile flavor, to the irritation of its Romanian member. Despite high rhetoric, it never had much political influence; its constituent peasant parties either failed to gain power in their respective countries or, in the few cases where they did so, became absorbed in the desperate but vain pursuit of purely domestic solutions to area-wide problems. In or out of power, these parties were quite nationalistic. The one authentic internationalist exception was the

Bulgarian peasant leader Aleksandŭr Stamboliski, and he was soon murdered for his pains by domestic supernationalists. The Green International's particular irrelevance and peasantism's general inadequacy were later exposed by the Great Depression.

5

The Communist parties, which came to power after World War II thanks to the Soviet army's conquest of the area, were politically weak during the interwar era. Though they might attract many genuine idealists, though their cadres usually bore persecution with courage, and though they benefited from the irascible habit of many local regimes of labeling all opposition as communistic, these parties were often discredited by their "antinational" identification with Russia (perceived as a historical foe and potential threat in Poland, Romania, and the Baltic States), the local ethnic minorities (popularly suspected as subversive), and atheism (especially damaging in the Roman Catholic countries). They were also hampered by their often inappropriate and vacillating approaches to issues of territorial revisionism, to the agrarian question, and to the peasantry as a class (was it a solid bloc or internally differentiated?)—weaknesses that, in turn, flowed from their organizational, financial, and moral dependence on the Comintern. Occasional lapses from slavish imitation of the caprices of the Moscow party line resulted in drastic purges, which racked the Polish, Yugoslav, and Bulgarian Communist parties with particular severity. Hence zombie-like obedience and the ritual discharge of assigned tasks became both a necessity for survival and a kind of psychological compensation for the Communist cadres' lack of real political influence.

The Communists, though, could capitalize on the pervasive discontent with poverty and oppression, the peasants' resentful alienation from the bureaucratic state apparatus, and the related failure of the peasant parties and leaders. Their appeals to social justice and revolution, while eliciting no immediate response, sustained an awareness of the Communists as representing a political alternative. Under Nazi occupation in World War II, the Communists finally enjoyed the advantage of long experience at underground organization, survival, and action. They were also able to capitalize on the fact that fear of Communism propelled many other local political actors, including even some original resistance movements, into compromising themselves by collaborating with the Axis occupier. Ultimately, however, the Communists' conquest of political power in East Central Europe at the close of World War II was determined less by local factors than by the decisive intervention of the Soviet Union. The two exceptions were Yugoslavia and Albania, where the Communists fought independently and won revolutionary national and civil wars.

6

Forms and styles of governmental activity—some of which persist to this day—passed through several similar sequences in the interwar East Central European countries. Yet throughout these changes and phases, the bureaucratic political class formed the effective and, except in Czechoslovakia and Hungary, virtually autonomous ruling class. Both its civilian and its military components were recruited from the so-called intelligentsia, which was identified simply by its possession of academic diplomas. The intelligentsia and, through it, the bureaucracy might be descended from the gentry, the middle class, or the peasantry; the peasant members tended to sever their cultural and behavioral ties with the village, despite sentimental and propagandistic professions of attachment. It might rule in association with the landed and entrepreneurial classes, but it was never a mere tool of the aristocracy or bourgeoisie. Universal suffrage did not protect the peasant masses from the intimidation or manipulation of this political class of bureaucrats and intelligentsia, and election results generally reflected its overall priorities, if not always its particular preferences. Indeed, in the area's more primitive regions, universal suffrage functioned as the bureaucracy's tool for breaking the traditional power of "feudal" notables over their peasant clienteles.

The tenacious but essentially stagnant power of this bureaucratic class largely accounted for the peculiar syndrome of immobility and instability in interwar East Central European politics—a syndrome to which, as has been indicated, Great Power pressures and ethnic tensions also contributed. Changes in cabinets were frequent; in constitutions, occasional. But the fundamental political reflections of social transitions were suffocated by these bureaucracies, and the social transitions themselves were often deliberately slowed. When the resulting tensions, aggravated by the Great Depression, became so acute as to erode the reliability of the parliamentary regimes of the 1920s as shields for the bureaucracy's ongoing power, it initiated or endorsed *coups d'état* that replaced the old regimes with royal, military, or political dictatorships or semidictatorships. The exception to this trend was Czechoslovakia, the area's economically most mature society.

The shift from parliamentary to authoritarian institutions in East Central Europe in the 1930s was also faciliated and supposedly vindicated by the impressive performance of the Great Power dictatorships, especially Nazi Germany, in energizing their economies and consolidating their societies. Over the great esteem in which German culture had traditionally been held in East Central Europe was now superimposed a new fascination—grudging or enthusiastic, but always respectful—with the Nazi political model. The imposing domestic and diplomatic successes of the Nazis, which contrasted so vividly with the apparent stagnation and decadence of France, gave the impression that authoritarian dictatorship was the wave of the future. States of lesser power, especially new or restored states, generally take as their model the political institutions and values of the seemingly strongest and most successful Great Power of the day. On the morrow of World War I, it appeared to be France; after the Depression, it became Germany. (And after World War II, it was to be the Soviet Union for a substantial portion of intellectuals and of the intelligentsia.) Furthermore, and with specific reference to East Central Europe, Nazi Germany's policies rendered territorial revisionism realistically "thinkable," and ethnic xenophobia, especially anti-Semitism, psychologically "respectable."

But the East Central European dictatorships of the 1930s would not or could not emulate the totalitarian dynamism of Hitler's example. Their commitments were essentially bureaucratic and conservative, at most technocratic and oligarchic. Projecting no mass ideology, they either failed or refused to elicit mass support. Despite their sonorous rhetoric of "the strong hand," they proved to be petty, brittle, often irresolute, and generally demoralizing.

Various Radical Right movements, drawing their political élan even more emphatically from the Nazi example, atavistic in their ideology but modern in their methods, claimed to supply the dynamism, the commitment to radical change, and the capacity to mobilize the masses that the local authoritarian regimes lacked or spurned. Noisiest in the countries with prominent and vulnerable Jewish minorities, the Radical Right leaders, while themselves usually educated and urbanized, appealed to the supposedly primitive, instinctive, and healthy revulsion of the peasant and proletarian "folk-masses" toward the allegedly decadent, "Judaized," and secular culture of their bureaucratic and bourgeois exploiters.

Indeed, the appeal and the appeals of Radical Rightism nicely reflected the condition of interwar East Central Europe as an agricultural society in a crisis of transition and fragmentation. Though not yet sufficiently developed and integrated to have moved beyond this demagoguery, this society no longer was stable and patriarchal enough to remain immune to it. The local Radical Right movements were, however, inhibited in their political offensives by the very fact that the authoritarian regimes that they sought to challenge already embodied a number of the Radicals' own professed ideological values; that is, they were undemocratic, ultranationalistic, and militaristic, and often mouthed Radical Right rhetoric even while repelling Radical Right bids for power. An even greater irony was that Hitler's regime, eager to extract maximum economic resources from East Central Europe for its own projected war effort, eventually endorsed the local forces of order and rationality—the authoritarian governments—against the counterproductive, albeit ideologically closer, enthusiasts of turmoil and upheaval—the Radical Rightists.

7

To the extent that the area's dictatorships scored any permanent successes, it was in the limited, though important, area of etatist economic investment, which did not, however, extend into radical social change or political mobilization. In all countries except Poland and Czechoslovakia, this was achieved through assistance from Nazi Germany in the form of bilateral exchange of local agricultural surpluses and raw materials for industrial equipment, investments, and technical support. While such economic aid was scarcely altruistic and was clearly designed to achieve regional hegemony and supplement the Reich's war economy, neither was it utterly exploitative or negative. Contrary to frequent allegations at the time and since, Germany did not flood East Central Europe with cuckoo clocks, aspirin, and thermometers in exchange for grains, minerals, and timber; rather, it supplied capital goods for industry, encouraged the diversification of vulnerable one-crop agricultures, and offered a steady market at reasonable prices. Nazi Germany's economic policy and behavior thus effectively supplemented its ideological, political, military, and diplomatic prowess in attracting Danubian and Balkan Europe to itself in the second half of the 1930s. Though the Serbs recoiled at the last moment, the Yugoslav governments had also climbed on this bandwagon.

In Poland, the equivalent etatist economic success was scored in the late 1930s without German assistance, through enforced local savings and investments. Czechoslovakia, in turn—or, more precisely, its western provinces of Bohemia and Moravia-Silesia—had already reached a substantial industrial plateau and made no analogous economic leap.

Though promising and perhaps indispensable, these etatist investment successes of the late 1930s were too little and too late to absorb more than a puny fraction of the surplus rural population into industrial employment or to spark self-sustaining economic growth. Hence they failed to transform interwar East Central Europe from an area of low industrialization, poor urban–rural imbalances, acute shortages of capital, and chronic agricultural poverty. Furthermore, as these etatist policies were generally predicated on attitudes of economic nationalism, they aggravated the understandable but nevertheless irrational craving for autarky and thus contributed to the fading of the fragile flower of intraregional cooperation, already withered by the arid winds of ethnic and irredentist enmity. Yet despite all its flaws and failures, this exercise in etatism in the late 1930s allowed for the accumulation of useful experience, which would be applied to the rapid, albeit mismanaged, regional industrialization drives of the post–World War II era. It is a thread of continuity from the interwar period into contemporary East Central Europe.

The searing trauma of the Great Depression had, of course, been the catalyst of the twin decisions to industrialize and to pursue beggar-thy-neighbor economic policies during the 1930s. It is difficult, even in retrospect, to appreciate and impossible to exaggerate the tremendous impact of this experience on the peoples and governments of interwar East Central Europe. As the world prices of agricultural commodities fell earlier and more steeply and remained longer at deeper troughs than the prices of industrial products, the Depression taught the dire lesson of the economic impotence of the agricultural-exporting countries relative to the industrial ones. Particularly hard hit were those extensive areas of East Central Europe that practiced the exclusive cultivation of grain for export, for grains are a commodity for which demand is notoriously inelastic, while supply is highly variable. Hence as prices dropped, the peasants desperately sought to compensate by increasing production, thereby merely further depressing prices—to their own impoverishment. In combination with governmental protection of infant native industries, with absurdly deflationary fiscal and monetary policies, and with exorbitant indirect taxes on such necessities as salt, matches, and kerosene (purchasable only from state monopolies), this price trend put virtually all industrial commodities—plows for production as well as textiles for consumption—out of the reach of the peasants and pushed them into bare subsistence and often into outright starvation. Economic despair then prompted political radicalization, which, in interaction with the judiciously orchestrated German drive for hegemony in the area, reopened the whole question of East Central Europe's international, domestic, political, and economic order. And on no occasion did the area's agrarian countries negotiate as a bloc with any industrial grain-importing country—an illustration of the ineffectiveness of the Green International.

The agricultural price disaster was paralleled and compounded by the West's abrupt, and probably unnecessary, withdrawal of all its capital credits to East Central Europe in the midst of the Depression. Unnecessary because while the sums involved were critical for the area's stability (even though they had often been applied unwisely), they were a relatively small fraction of the Western creditors' total international investments. Industrial output, capital formation, and employment fell precipitously, with calamitous political repercussions. This politico-economic myopia of the West, which had been foreshadowed by France's persistent refusal to support its alliances in the area with adequate trade relations, virtually invited Nazi German penetration. Germany, in turn, did not intend to integrate East Central Europe into the world economy, but the reverse: Germany wished to tie it to its own economy and thus create an autarkic *Grossraumwirtschaft*, supplementing and facilitating its projected political and military conquest of Europe. The Soviet Union would pursue an analogous policy after World War II.

A particularly powerful instrument of this economic strategy was the blocked-currency device, whereby the high sums paid by Germany for its huge purchases of agricultural goods and raw materials from East Central European countries were held in blocked accounts at the Reichsbank and could be "cleared" only by East Central European purchases of German commodities. Though the local governments sometimes grumbled about being obliged to take German equipment when they would have preferred being paid in convertible currencies, on balance they appreciated being rescued by Berlin from the economic and political disaster of otherwise unsalable agricultural surpluses. Nazi Germany thus acquired control over the region's economy by first dominating its exports, then (through exports) controlling its imports, and finally rendering it utterly dependent on continuing German purchases, supplies, spare parts, and infrastructure. In this way, Germany achieved a position approaching both monopsony and monopoly. By 1939, on the eve of World War II, Germany's economic hegemony over East Central Europe was more categorical than it had been in 1913, demonstrating that the political advantages that accrued to it from the replacement of the Habsburg Empire by several smaller states were paralleled by economic opportunities.

Thus the combination of Nazi Germany's ideological, diplomatic, political, and economic drives paved the way for its military conquests. In one form or another, all the states of the region eventually succumbed to German offensives, as resisting victims (Poland, Yugoslavia), as passive victims (Czechoslovakia), as calculating satellites (Hungary, Romania, Bulgaria), as ephemerally "independent" dependencies (Slovakia, Croatia), or as trophies from the midwar collapse of Italy (Albania).

8

If the preceding discussion has emphasized interwar East Central Europe's internal weaknesses and external vulnerabilities, and hence appears to signal a negative judgment, this would be an erroneous impression of the author's intention and ultimate conclusion. That impression arises partly from the fact that the most positive political achievement of the region's states during the interwar era is so obvious as easily to pass notice: they legitimated their sovereign existence in the world's eyes beyond Nazi or Stalinist capacity to obliterate. (The three Baltic States are an exception, but even they are granted distinct republican status within the Soviet Union.) Thus contemporary East Central European Communist historians, otherwise highly critical of their countries' interwar social and economic policies, join the "bourgeois" émigré scholars and politicians in valuing highly the sheer fact of interwar state independence and judging it as a historic advance over the region's pre–World War I political status. (Again, the Baltic States are treated as a negative exception.) No Communist, Soviet or local, would any longer indulge in Molotov's contemptuous dismissal of interwar Poland as "this ugly offspring of the Versailles Treaty" (speech of October 31, 1939). Nor do respectable German writers repeat their interwar predecessors' persistent derision of the alleged Polish *Saisonstaat* or the Czechoslovak *staatliches Missgebilde*. Thus despite major and avoidable failings (too little intraregional solidarity, too much overpoliticization of human relations, too little strategic government intervention in the economy, too much petty government interference with the society), thanks to the political performance of East Central Europe during the interwar era, it is impossible today to conceive of the region without its at least formally independent states. In retrospect, one must assign greater responsibility for the catastrophes suffered between 1938 and 1941 to the malevolence, indifference, or incompetence of the Great Powers than to the admittedly costly mistakes of the East Central European states.

We have now reached a point where it is appropriate and imperative to attempt a political analysis—not a military survey—of what happened in East Central Europe during and immediately after World War II, a war that replaced Germany with the Soviet Union as the region's hegemonic power and thereby precipitated profound political and socioeconomic transformations in all its states.

2

World War II

1

It has been provocatively argued that in regard to European political history, World War II was not a discrete crisis, but the third act in a continuous tragedy whose two earlier acts consisted of World War I and the interwar failure of will to defend democratic and humane institutions against totalitarian challenges. Hence, so this argument runs, the period from 1914 through 1945 might appropriately be designated as Europe's second Thirty Years' War—with, however, a totally different global outcome from the first Thirty Years' War of the seventeenth century. From that earlier crisis, Europe had emerged as the complete master of its own political destiny, despite all its internal political fragmentations and frictions. Even the occasional flirtations by one or another of its constituent members with a culturally non-European power, such as France's with the Ottoman Empire in the baroque era, did not compromise Europe's capacity to regulate and control its own fate and fortunes. World War I, however, yielded only an internal European stalemate whose breaking required the decisive economic and military intervention of the United States and of the extra-European dominions of the British Empire. The failure in the interwar years to restore an effective intra-European balance of power, and then World War II, only ratified irrevocably this political collapse of Europe as an autonomous actor and the transfer of control over its destiny to the non-European superpowers.[1]

Whatever the merits of this argument in a global sense—and even here it is quite problematic—as regards perceptions in and of East

Central Europe, its blurring of the different outcomes of the two world wars is quite dubious. In dismissing the macropolitical, or balance of power, significance of the interwar independence of the several states of East Central Europe, it implicitly invites the error alluded to in the closing section of Chapter 1, the error of underestimating the moral value of that very independence—a moral value that is today being vigorously reasserted. And it was precisely this error that compromised and undermined Nazi Germany's ability to consolidate the hegemony over this area that it had achieved between 1938 and 1941. Had Hitler's "new order" not been such a blatantly transparent screen for racial imperialism, exploitation, and genocide, had he instead offered a dignified status and role to the East Central European peoples under his hegemony, he might have averted their resistance movements, harnessed sustained local support, and possibly consolidated his East Central European conquests as a rampart against the Soviet Union. But the defeat of Hitler's Germany did not prompt the recovery of effective East Central European political independence. Though the formal sovereignty of the states that had emerged from World War I was juridically restored after World War II—a significant tribute to their legitimacy—political control over them was transferred to the Soviet Union, whose rulers subjected the region to a profound socioeconomic revolution. To a substantial degree, the implementation of that revolution was facilitated by the wartime experiences of the East Central European societies and especially by the traumas inflicted on certain sectors of those societies under Nazi German occupation. Thus a certain developmental and even causal continuity exists between the wartime years of German conquest and the immediate postwar years of early Communization, and a profound break separates this double period from the earlier interwar decades, which, in retrospect, refuse to merge into the one continuous European crisis stipulated by the theory cited earlier. Rather, they stand out in sharp relief against the post-1938 picture of sequential German and Soviet hegemonies.

2

Under the provisions of the German-Soviet Pact of August 23, 1939, as modified on September 28, 1939, Nazi Germany acquired the territory that included most of Poland's industry and ethnically Polish population, while the Soviet Union's share of this partition contained the oil fields and the Belorussian and Ukrainian populations, together with a substantial minority of Poles. Though the Germans were to overrun and rule this Soviet zone of Poland between 1941 and 1944, the Soviet Union would reannex the bulk of it after the fortunes of war had turned, compensating Poland in the west at Germany's expense with most of Pomerania, much of Brandenburg, and all of Poznania and Silesia (the Oder-Neisse, or Odra-Nysa, Line). East Prussia would be shared between postwar Poland and the Soviet Union.

Immediately after the implementation of this pact of 1939, through the dual German and Soviet invasions and partition of Poland, Nazi Germany annexed to the Reich that portion of interwar Poland that before 1919 had been part of Prussia, some segments that had briefly belonged to Prussia between 1795 (Third Partition of Poland) and 1807 (Treaty of Tilsit), and parts of central and southern Poland that had never been under Prussian/German rule. The rump of the Nazi share of Poland—the territory between these expanded Reich frontiers and the German–Soviet demarcation line—was now termed the *Generalgouvernement*—without any eponymous national identification in an effort to erase the very name of Poland from the map of Europe.[2] Poles and Jews from the areas annexed directly to the Reich were deported en masse into the *Generalgouvernement*, and industrial equipment was extracted from it in a deliberate German policy of inflicting both overpopulation and economic spoliation. Eventually, Polish workers were deported into the Reich as slave laborers, while Jews were transported from all over Europe to the several extermination camps that were erected in the *Generalgouvernement*. While the Nazi intention was to exterminate all Jews and Gypsies, Poles in general were to be reduced to "mere" peonage, but the Polish intelligentsia was also intended to be physically liquidated. After the Battle of Stalingrad, Governor General Hans Frank suggested a belated effort to court the Poles as possible auxiliaries against the now steadily counterattacking Soviet armies, but Himmler and Hitler vetoed his proposal. It would, in any event, not have been taken up by the Poles, who were by then irrevocably alienated by the Nazi excesses and atrocities.

Building on a traditional familiarity with conspiracy that dates back to the era of the original partitions, Polish society managed to develop a highly effective underground army and state apparatus, operative not only in the *Generalgouvernement*, but throughout most of the territory of the interwar republic. With the possible exception of Yugoslavia, no country resisted the German occupation as tenaciously or suffered as savagely as did Poland. Six million of its citizens—half of them Jews—were killed, one-third of its houses were destroyed (nine-tenths in Warsaw), and its social infrastructural investments of the interwar decades were ruined. And the achievements of the Polish resistance movement were indeed prodigious. It tied down approximately 500,000 German occupation troops and, according to official German figures, prevented one out of every eight *Wehrmacht* transports headed for the Russian front from reaching its destination.[3] The climax of the Polish resistance effort was the heroic but abortive Warsaw insurrection of August 1 to October 2, 1944. Jewish uprisings had also been launched and repressed earlier in the ghettos of Warsaw (April 19–May 16, 1943) and of Białystok and Wilno (September 1943). Abroad, substantial Polish military units fought against the Germans on most Allied fronts: Norway, France, the Battle of Britain, North Africa, Italy, Normandy, the Lower Rhine, and the Soviet Union. And Poland was the only Axis-occupied country in Europe without a quisling.

In the Soviet half of interwar Poland, which Moscow was to control from September 1939 until the German advances of the summer of 1941 and then uninterruptedly since the spring of 1944, the authorities were more careful than were the Germans to be formally attentive to the principle of national self-determination. First, they waited until September 17, 1939, to invade it, after the German armies—which had launched their attack on Poland on September 1—had manifestly won the campaign. Moscow thus demonstratively evaded the charge of stabbing a still viable Polish defense in the back. Second, the Soviets took pains to proclaim that their only interest in occupying eastern Poland was to protect and redeem fraternal Ukrainian and Belorussian populations, who did indeed constitute the regional majorities. Third, they took the trouble to arrange plebiscitary elections on October 22 for local Ukrainian and Belorussian "national assemblies," which a week later ostentatiously petitioned for the admission of their regions into the Soviet Union—petitions that the Supreme Soviet graciously granted on November 1 and 2, respectively, simultaneously conferring Soviet citizenship on all the inhabitants. Fourth, the Soviet authorities fastidiously transferred the Wilno (Vilnius, Vilna) district to Lithuania, which was still independent, as ostensible evidence of their supposedly disinterested benevolence.

Underneath this benign veneer, Soviet policy was in practice harsh and even brutal. Private productive property was socialized and land collectivized; ethnic Poles were extruded from public and professional life; and between 1.25 and 1.5 million people were deported to Siberia, the Russian Arctic, and Soviet Central Asia, where approximately half of them died. This Soviet repression was targeted particularly against the local Polish elite and intelligentsia—gentry, professional classes, clergy, civil servants, and interned military officers.

Having declared in September 1939 that the Polish state had ceased to exist, Moscow logically refused to recognize the Polish government-in-exile, which was established after that month's military fiasco, first in Paris and then, after the fall of France, in London. The government-in-exile was in continual liaison with and endorsed by the underground resistance movement in Poland. It formally reserved the rights of the Polish state to all its prewar territory. Rather ominously but pragmatically, the British government, though recognizing and eventually hosting the Polish government-in-exile, distanced itself from the Polish protests against the Soviet annexation of eastern Poland. As early as October 26, 1939—that is, even before Hitler's invasion of the Soviet Union—Lord Halifax noted that the Soviet territorial claims corresponded substantially with those that Lord Curzon had deemed appropriate in a border recommendation that this earlier British foreign secretary had given in 1920. Though Churchill would try to reconcile the Polish and Soviet governments to each other at intervals throughout the war, Britain never budged from this position, which indicated its readiness to acquiesce in the mutilation of its Polish ally for the sake of first gaining and then keeping the more powerful Soviet ally against Nazi Germany.

On July 30, 1941, after the Nazi German invasion of the Soviet Union on June 22, the Polish and Soviet governments, now at war against the same enemy, did resume diplomatic relations—but without warmth. Though the Soviets forsook their pact of August 1939 with Nazi Germany, which had initiated the September 1939 partition of Poland, they refused to commit themselves to the restoration of the prewar frontier, even while their own military fortunes were very low. During the *Wehrmacht*'s initial deep penetration into the Soviet Union, far eastward of any plausible Polish–Soviet boundary, the Polish government-in-exile assented to leaving this issue unresolved and devoted itself to implementing Stalin's agreement that it be permitted to raise an army from among those Polish military units that had fallen into Soviet captivity in September 1939. But even this effort at incipient military cooperation soon turned bitter. Most of the Polish officers who were now released to form the cadre of the new army were resentful over and emaciated by their treatment during the preceding two years. More ominously, many others, known to have fallen into Soviet hands alive and well, now failed to reappear. The Polish units that were formed accused the Soviet authorities of starving them of equipment and rations, while the Soviets retorted that such charges were a specious Polish alibi to avoid combat. In August 1942, after ceaseless friction with its Soviet hosts, the Polish army was evacuated from the Soviet Union via Iran to the Middle East, eventually to participate very impressively in the Western Allies' Italian campaign. The Soviets, in turn, accused it of deserting the eastern front on the eve of the critical Battle of Stalingrad. In fact, its departure for the Mediterranean theater was probably quite welcome to Stalin, as this move eliminated any possibility that Poland might be liberated by an army not under complete Soviet control.

The sinister implication of the non-reappearance of many Polish officers who had been captured by the Soviets in 1939 was soon concretized and led to a new rupture in the recently restored and always fragile Polish–Soviet relationship. On April 13, 1943, the Germans announced the discovery of the mass graves of several thousand Polish officers in the forest of Katyn, just west of Smolensk, and offered evidence that they had been shot before the *Wehrmacht* occupied this area in the summer of 1941. On April 26, after the Polish government-in-exile requested the International Red Cross to examine the site and the evidence, the Soviet Union severed diplomatic relations, charging the "London Poles" of acting in collusion with Hitler. Though neither the Soviet nor the postwar Polish Communist authorities have ever conceded it, the evidence overwhelmingly points to the Soviet secret police, or NKVD, as the murderers of the 15,000 Polish officers and soldiers.[4] To this day, the unacknowledged crime of Katyn burdens Polish attitudes toward the Soviet Union and poisons the relationship between the Polish people and their government.

It was, of course, highly convenient to the Soviet government to be disencumbered of relations with, and hence of obligations to, the Polish government-in-exile as the advancing Soviet armies approached Poland. In effect, this gave Stalin a virtually free hand to make his own *de novo* political arrangements on the ground. He prepared for this opportunity in several ways. First, after the evacuation from the Soviet Union of the original Polish army raised for the eastern front, he recruited a second army from among the remaining Polish captives of the 1939 campaign, commanded by a few amenable officers of the interwar Polish army but mainly by Soviet officers of Polish ethnic origin. Modeled after the Soviet army, with its political commissars and the like, this so-called Kościuszko Division became the nucleus of the postwar Polish army, and the anniversary of the day on which it first went into combat, October 12, 1943, at Lenino, is still observed as Army Day in Poland. As it advanced westward into Poland with the Soviet armies, this division was repeatedly expanded into corps and then into armies through mergers with Communist partisan bands and the enforced recruitment of units and individuals from the general resistance movement, which, though loyal to London, was crippled by the absence of any Western military presence in Poland to match and check the Soviet one. Second, Stalin caused a Polish Communist underground movement to be organized during the German occupation in competition with the initially far larger, more authentically nationalist, and more effective London-affiliated general resistance movement known as the Home Army. Third, he organized the nucleus of an eventual alternative Polish government from among Communist functionaries and leftist intellectuals resident in the Soviet Union and now politically schooled by his own apparatus. It was termed the Union of Polish Patriots.

When the Soviet armies crossed the pre-1939 Polish–Soviet frontier in January 1944, the Soviet government denied that any valid international border had been crossed and insisted that its forces were still operating within the territory of the Soviet Union. Only in July 1944, when its forces crossed the Bug River—designated as a suitable border by both the Curzon line of 1920 and the German–Soviet partition line of 1939—did Moscow concede that they had entered Poland. It then promptly unveiled the Polish Committee of National Liberation, composed of Communists who had been active in the underground during the German occupation, activists from the Union of Polish Patriots, and splinters from the Socialist and peasant movements who, though non-Communist, had come to the conclusion that it would be futile and quixotic to resist the Soviet presence in their country. The Soviet army transferred civilian administrative responsibility in the Polish areas that it liberated to the Committee of National Liberation, which on January 1, 1945, designated itself—with Moscow's blessing—the Provisional Government of Poland. Meanwhile, units of the far more substantial anti-German resistance movement that remained loyal to the government-in-exile in London were arrested by the Soviets and executed, exiled, or forcibly incorporated into the Polish army that was being built from the Kościuszko Division. The Soviets also did not exert themselves to save the London-loyal Warsaw insurrection of the summer of 1944 from being suppressed by the Germans with much loss of Polish life and property, thus further decimating the old Polish elite. They understood that this insurrection was intended to liberate the capital by the national resistance movement rather than by the Soviet forces, who might then be received as guests rather than as conquerors. Hence their readiness to see it snuffed out.

Though in hindsight this process of the destruction of the anti-German and anti-Communist Polish national resistance movement is often described as though its outcome was foreordained, it was anything but smooth. Indeed, despite the overwhelming and decisive Soviet presence, Poland was wracked by a real civil war, lasting well into 1947 at least, in which the Soviet-backed Communist forces, the surviving national resistance cells, and Ukrainian partisans fought one another with great ferocity and desperation.

As World War II approached its close in Europe and again shortly after its conclusion, the Big Three leaders of the Allies confirmed the political fate of Poland at the Yalta and Potsdam conferences of February 4 to 11 and July 17 to August 2, 1945. Stalin's claims concerning the Polish–Soviet frontier were now ratified by the Western Big Two; Poland was compensated for its lost eastern lands by the acquisition of the German territories up to the Oder-Neisse Line and southern East Prussia. To make room for the Polish population that would be extruded from the eastern lands, the German population was to be moved westward out of the newly acquired Polish western region. The Soviet-sponsored Provisional Government of Poland was endorsed by the British and the Americans—who thus sacrificed their faithful, if somewhat stubborn, Polish allies in exile—on condition that it incorporate a few "London" Poles as individuals and that it nominally commit itself to early and free elections in Poland—a provision that would mean whatever Stalin might choose to have it mean. Thus a war that had begun to preserve Poland's authentic independence from Nazi Germany ended with its being doubly dependent on Soviet Russia: Poland was to be governed by a cadre determined to match its social, economic, and political life to the Soviet model; Poland's international security was to be entirely dependent on Soviet protection of its new western frontier against future German revanchism.

3

The wartime behavior and experiences of the Czechoslovaks was in several ways the direct opposite of the Poles' patterns, yet their fate since World War II is similar. Whereas the Poles fought the Germans in 1939 despite catastrophically unfavorable odds, the Czechoslovak regime capitulated in 1938, though its odds were not as poor. Whereas the Poles then resisted the occupation and suffered enormous human and material losses during the war, the Czechs were largely quiescent and benefited from the German industrial plants being moved into their land and thus out of the reach of the British and American bombings. Not that the Czechs were enthusiastic collaborators with the Nazis, whom they indeed resented. They simply kept a pragmatically low profile and avoided the risks of resistance and reprisals. By and large, only Czech Jews and intellectuals suffered persecution. Whereas the Polish government-in-exile provoked Stalin's wrath by opposing—perhaps imprudently but certainly bravely—his territorial and political demands, its Czechoslovak counterpart, also based in London, toadied to him. To no avail. Czechoslovakia was ultimately integrated into the Soviet security, political, and socioeconomic systems at least as firmly as was Poland, though with slightly later timing.

Czechoslovakia entered the pre-Munich crisis in the summer of 1938 with some strong domestic cards, but its leaders never played them. Constitutionally and politically, the government was solidly anchored in a democratically elected parliament and in Czech public support. Admittedly, the leaders of the Sudeten German minority were by then openly seditious and the loyalty of many Slovak leaders was dubious, but there is no doubt that the Czech nation—the country's dominant majority—faced the critical summer of 1938 resolute and confident. Moreover, the military establishment was thoroughly competent and professional. Indeed, during World War II, Hitler once noted that during the 1930s only two European states, his own Reich and Czechoslovakia, had really seriously prepared for war, and at the postwar Nürnberg war-crimes trials, Field Marshals Wilhelm Keitel and Erich von Manstein testified that in 1938 the Czechoslovak fortifications could have offered formidable resistance to the *Wehrmacht*. Given its naturally defensible and well-fortified frontiers, its technologically advanced armaments industry, and its disciplined and literate population, Czechoslovakia's potential military position in September 1938 was not as apparently hopeless as Poland's after the German-Soviet Pact a year later. Hence the capitulation of President Edvard Beneš to Munich, for which he never accepted responsibility but blamed the Great Powers exclusively, was not a rational calculation of military and political odds, but a profound failure of political and psychological nerves.

The point is that there are certain ultimate leadership decisions that determine the moral, even more than the material, fate of future generations, decisions that the leaders of even small states cannot "rationally" or "logically" abdicate to their Great Power patrons without compromising their own integrity.

In the aftermath of the Munich conference, Czechoslovakia suffered huge losses of territory and resources to Germany, Hungary, and Poland; but these losses were reversed at the end of World War II. Less remediable than the material damage was the psychological one, which would ultimately benefit the Soviet Union and the domestic Communists. The public's confidence in the prewar international system and in its own leaders was sapped; the elite's morale, broken. Even the shattering defeat at White Mountain in 1620, when battle had been accepted by the Czechs, was less demoralizing than this humiliating acquiescence to Munich in 1938. The last, but scarcely the least, of this episode's many hard lessons is that the sacrifice of Czechoslovakia did not save the peace.

Beneš resigned the presidency on October 5 and left his country on October 22, 1938. As the French government, unable to forgive the man whom it had betrayed, refused him any contacts, Beneš went on to exile in Britain, where the government also kept him at arm's length until well into World War II. His formal successor at home was the elderly and apolitical jurist Emil Hácha. Slovakia was granted extensive autonomy on October 6, as was Ruthenia, the country's easternmost province, two days later. The state's name was hyphenated to Czecho-Slovakia. The surviving political leaders drew the logical inference from Munich that, their rump state (also known as "the second republic") being henceforth utterly dependent on Hitler's benevolence, they had best offer him their willing collaboration. Accordingly, the constitution of 1920 was nullified, the Czech party system suspended, the Communists banned, the remaining German minority given privileged status, the Jews restricted, censorship extended, and democracy vilified in public propaganda. An extraterritorial road connecting Silesia and Austria was put at Germany's disposal, and the remaining heavy armaments were transferred to it. Finally, the new Czecho-Slovak foreign minister, František Chvalkovský, beseechingly promised full policy compliance with, and reliance on, Germany "if Germany will allow this."

For a brief period, it appeared that this obsequiousness might work. In the autumn arbitration proceedings concerning the new frontier with Hungary, for example, the German delegation was less vindictively hostile to the Czecho-Slovak case than was the Italian. The Germans had also initially backed that relatively moderate wing of the Slovak People's party that was prepared to accept autonomy within what was left of the general state, rather than the radicals who craved total Slovak independence. Indeed, since the Czecho-Slovak rump state was a true satellite and entirely dependent on the Reich, it would appear to have been in Berlin's interest to stabilize and sustain it.

Hitler, however, acting for reasons and from motives that remain somewhat unclear, chose otherwise. In mid-March 1939, he took advantage of an internal crisis between the Prague central government and the Slovak autonomous one to impose the Slovak radicals on the moderates and thus to elicit a declaration of full Slovak independence under German protection (de facto vis-à-vis Hungary). Simultaneously, he utilized Hácha's suppliant visit to Berlin to browbeat the old man into accepting German military occupation of, and the politico-administrative German Protectorate over, the rump Czech lands of Bohemia and Moravia. Operationally, the military occupation was a mere police action, all effective Czech defenses having been surrendered after Munich and any (unlikely) urge to offer quixotic resistance at this point having just been allayed by Hácha's capitulation. Hungary, meanwhile, on being denied Slovakia, consoled itself by reannexing Ruthenia.

On balance, these frenzied actions of March 13 to 16, 1939, were a blunder on Hitler's part. Politically, he gained no greater control over the territories now under his formal protection than he had in fact enjoyed since Munich, while internationally, he finally aroused even the hitherto complacent British government from its illusions of appeasement. The German occupation of the Czech rump state on March 15, 1939, thus led directly to the British guarantee of March 31 to Poland, with consequences fateful for the world and fatal to Hitler and his Third Reich.

Unlike the *Generalgouvernement* for his Polish conquest, Hitler preserved the legal fiction of Czech autonomy in his Protectorate of Bohemia and Moravia. The Hácha puppet government was formally maintained, a small militia authorized, and a single mass political organization called the National Solidarity Movement tolerated. But the real locus of power was, of course, the office and apparatus of the supervisory Reich Protector, staffed by Germans.

Apart from its universal aim of exterminating the Jews, German policy toward the Protectorate's Czechs was to brutalize the intellectual and professional classes—from among whom came most of the 36,000 to 55,000 Czechs who were executed directly or who died in concentration camps during the war—and to coddle the workers, peasants, and artisans with full employment, ample rations, high wages, and steady purchases at good prices.[5] By and large, this policy of "bribery through the stomach" succeeded in rendering the Protectorate one of the most quiescent and productive parts of Axis-occupied Europe. Apart from the rather spectacular ambushing of the Acting Reich Protector, Reinhard Heydrich, on May 27, 1942 (he died of his wounds on June 4)—and even in this case, it is noteworthy that the escaped assassins, who had been sent from Britain, were later betrayed to the Gestapo by their Czech co-nationals—and the ostentatious but operationally insignificant uprising of Prague, on May 5 to 9, 1945, when the European war was virtually over and the Nazi regime had disintegrated (and which left the city virtually unscathed), the Czech resistance was rather minimal throughout the war and never enjoyed the sympathy, let alone the participation, of the lower classes. One stance on which the otherwise rivalrous Hácha puppet government inside the Protectorate and the government-in-exile that Beneš formed after the fall of France could agree was to avoid risks and damages while waiting for the ultimate fate of the country to be settled by the exertions of the Great Powers. Perhaps it is precisely because of this relative paucity of Czech resistance and suffering that the moral and psychological wounds of the occupation years have cut so deep.

Interestingly, active resistance was more vigorous in the nominally sovereign Slovak puppet state than in the Protectorate. Hoping to capitalize on the fact that Slovak nationalism was anti-Czech and anti-Hungarian but not anti-German, the Nazis initially aimed to make Slovakia into a showcase displaying to all Europe the rewards of collaborating with them. Then, once the war was on, Slovakia became an exemplar of a more general wartime policy of Nazi Germany—reluctantly sacrificing the more sympathetic but unruly local Radical Rightists for the sake of political stability and economic productivity, which in Slovakia was maintained by a regime of clerico-authoritarian conservatives. The bargain between the Slovak conservatives and the hegemonic Germans appeared to be mutually profitable: Slovakia's administrative and political autonomy was maintained, and its economy was expanded and modernized; in return, it supplied food, raw materials, and semiprocessed goods for the German war effort. Slovakia also collaborated in the roundup and hence the extermination of Jews. But as early as September 1939, there had been some mutinous behavior in protest against the regime's collaboration with the German invasion of Poland, a nation toward which Slovaks feel affinity, and by 1943 the bargain was souring somewhat in Slovak eyes as laborers were drafted to work in Reich industries and troops to fight on the eastern and Italian fronts.

After many sporadic yet small-scale acts of sabotage, desertion, shirking, hoarding, and evasion, the resistance—incorporating Communist as well as non-Communist elements—organized itself under a common political leadership and program in November 1943, but for symbolic effect called its decisions of that month the Christmas Program. It organized the Slovak National Council, called for the reestablishment of a common state with the Czechs, but this time with more equality to the Slovaks as a distinct nation, and invited this future, egalitarian Czechoslovakia to lean in foreign policy on the Soviet Union "as the protector of the freedom and universal progress of small nations in general and of Slav nations in particular."[6] Simultaneously, the Slovak National Council was in contact with disaffected officers in the puppet government's own army, with a view to arranging for a Slovak leap out of the Axis and into the Allied camp at an opportune moment.

The choosing of such a supposedly opportune moment for a volte-face was, of course, a highly delicate matter. No Slovak wished to repeat the unhappy experiences of Italy and Hungary. The first had surrendered prematurely to the Allies in September 1943, and the second was suspected by Hitler of intending to do so in March 1944; both were promptly invaded and subdued by vigorous German counterstrokes. In the event, the Slovaks' timing was forced by Romania's switch of sides on August 23 to 25, 1944 (see section 6), which elicited a preemptive German occupation of Slovakia on August 29, thereby sparking a Slovak uprising under the formal leadership of the relatively new Slovak National Council but commanded and partly manned by the officers and units of the puppet government's own army together with the council's guerrillas. The uprising managed to survive for two months in central Slovakia until it was finally overcome by the *Wehrmacht* in heavy fighting that lasted until October 28. It received only slightly more Soviet assistance than did the nearly simultaneous insurrection in Warsaw. Only in the first quarter of 1945, after the defeat of the rebels, did the Soviet army clear Slovakia of the Germans. Yet, in contrast to its Polish analogue, the Slovak uprising did achieve an important political success. Its tenacity and heroism forced the reluctant Beneš government-in-exile to assent to greater autonomy and equality for postwar Slovakia within Czechoslovakia than it had

had in the interwar republic. It is to the vicissitudes and maneuverings of Beneš that we now turn.

As mentioned earlier, the self-exiled Beneš was initially treated as a pariah by the governments of Britain and France, which hoped that they had purchased peace at Czechoslovakia's expense at Munich in September 1938. Even after the outbreak of the war a year later, Benes and his fellow Czechoslovak public figures in exile received British and French recognition only as a national committee, not a government. But after the fall of France and the ascent of the anti-appeasers Churchill and Anthony Eden to power in Britain, that country extended recognition to the exiles as the Provisional Czechoslovak Government. The adjective *provisional* rankled, as it placed the Czechoslovaks on a lower juridical plane than the other governments-in-exile from the German-occupied European countries, and was dropped after Hitler's invasion of the Soviet Union, when Churchill and Stalin simultaneously extended full de jure recognition to Beneš's government on July 18, 1941. Finally, after tenacious lobbying by Beneš, the British government was persuaded to repudiate the Munich agreement on August 5, 1942, to be followed in this gesture by De Gaulle's Free France movement on September 29—four years to the day after the infamous conference.

Beneš's juridical self-presentation then stood as follows: (1) the Czechoslovak Republic, founded in 1918, continues to exist uninterruptedly in the legal personality of his exile government; (2) his resignation from the presidency after Munich was legally invalid, and he never ceased to be the president of Czechoslovakia; (3) neither the Protectorate nor the Slovak secessionist state has legal validity; (4) the Munich agreement was invalid from the beginning, and not merely after the Germans violated it by occupying the rump Czech state in March 1939; (5) the territorial losses to Germany, Hungary, and Poland that were imposed on Czechoslovakia immediately after Munich are therefore also invalid.

The preceding paragraphs give some indication of Beneš's prodigious talents as a negotiator and a casuist. These traits of intellectual self-assurance, of persistence, of high confidence in his ability to spin legal and rhetorical formulas to paper over political issues were also revealed in his wartime behavior toward his fellow exiles and toward the Big Three leaders. Within his own Czechoslovak government-in-exile, Beneš systematically destroyed every person of independent judgment, until he was accountable to no one and controlled all organs and policies. He suppressed the exile representations of those Czech and Slovak parties that he deemed to have been contaminated by Munich. The handful of Sudeten German democratic politicans who, at great risk and with much courage, had defied the Nazis were treated shabbily. Toward Stalin, on the contrary, Beneš's behavior was simultaneously politically fawning and intellectually condescending. By the summer of 1943, with the British and American armies still bogged down in Italy and the Soviet ones relentlessly advancing, Beneš astutely anticipated that East Central Europe would be liberated by the Soviets and accordingly decided to ingratiate himself with Stalin and to tutor him on the true Soviet interest in the region. He quickly distanced himself from the neighboring London Poles, with whom a year earlier he had anticipated forming a postwar confederation, but who had became anathema to Moscow after the exposure of the Katyn forest massacre, and he invited the Communists into his government-in-exile. Then he requested the special Treaty of Friendship, Mutual Aid, and Postwar Cooperation with the Soviet Union (the first one between a small Allied state and that Great Power) and, against the advice of his British hosts, flew to Moscow in December 1943 to sign it. On this occasion, Beneš saw fit to advise Stalin and Molotov that they should extirpate "feudalism" in Poland and Hungary and to denigrate the Romanians and Yugoslavs.[7] A year later, he succumbed supinely to a demand by Stalin that he cede to the Soviet Union the easternmost Ruthenian province of interwar Czechoslovakia (also termed the Carpatho-Ukraine). Though very poor, it is strategically important because it controls several Carpathian mountain passes giving access from the Ukraine into the Hungarian plain.

Beneš even had the intellectual conceit to elaborate a pseudo-profound, semisociological "theory" to rationalize his pragmatic calculation that only through such a posture of flattery of and submission to the Soviet rulers would he be enabled to establish his own government in Czechoslovakia at war's end and would he be spared Communist criticism as a "Munich poltroon." This theory had it that the Western and the Soviet societies were on convergent tracks, with the former progressing from laissez-faire capitalism toward welfare-state socialism, and the latter evolving from totalitarianism toward social democracy. Czechoslovakia under Beneš's government should facilitate this pair of healthy sociopolitical trends by serving as a postwar bridge between the British and Americans and the Soviets.[8] Though he was indeed permitted—unlike the London Poles—to bring his government-in-exile home at war's close (albeit with much expanded Communist participation), Beneš was not destined to be spared the Soviet-sponsored Communist subversion of his authority three years later.

4

Though the Hungarians are probably the most Anglophile nation of East Central Europe, they served in World War II as one of Hitler's calculating satellites. The reason for this seeming anomaly was their passionate irredentism. Interwar Hungary was the main loser of the Paris treaties system, which had closed World War I, being truncated to only one-third of its historic territory, two-fifths of its total prewar population, and two-thirds of its Magyar people. The accompanying loss of natural and economic resources was also staggering. Zealous revisionism, directed against Czechoslovakia, Romania, and Yugoslavia—the leading beneficiaries of the mutilation inflicted on Hungary—was the general, well-nigh universal, response of interwar Hungarian society to these harsh terms. By the eve of World War II, Hitler and Mussolini were available as Great Power champions for Hungary's grievances, and under their patronage it soon achieved partial but substantial satisfaction in four installments: (1) in the aftermath of Munich, Hungary recovered a strip of southern Slovakia and southwestern Ruthenia; (2) when Hitler imposed his Protectorate on Bohemia and Moravia and midwifed the formal independence of Slovakia, he also permitted Hungary to reannex the rest of Ruthenia; (3) in August 1940, he forced Romania to return to Hungary northern and eastern Transylvania; (4) when he and Mussolini smashed and partitioned interwar Yugoslavia in April 1941, Hungary was enabled to reacquire a part of its earlier loss to that country. Thus thanks to its association with the Axis, interwar Hungary doubled in size between 1938 and 1941 (which, however, still left it far smaller than its historic extent). Again thanks to its association with the Axis, Hungary would eventually be obliged to relinquish all these territorial gains at the close of World War II.

Supping with the devil proverbially requires a long spoon, and the spoon of the Anglophile, whiggish, old-fashioned, liberal-conservative Hungarian ruling classes was not long enough to avoid paying a price for Hitler's patronage of their territorial expansions, though they maneuvered resourcefully to try to hold that price down. For starters, the lower classes were substantially weaned away from traditional Magyar Anglophilia, not only by Nazi Germany's sponsorship of Hungary's wartime territorial expansion, but also by its even earlier bulk purchases of Hungary's otherwise unmarketable agricultural produce and absorption of Hungary's surplus manpower as seasonal labor in the Reich. This genuine popularity of Nazi Germany among Hungary's lower classes was grist for the Radical Right mill of "ingratiation through imitation"—that is, the recommendation that Hungary ensure itself continuing and extended Nazi support for the full array of its territorial, political, and economic ambitions in the Carpatho-Danubian basin by coordinating its internal political institutions and processes (as well as foreign policy) ever more closely with those of Hitler's Germany.

Though the fastidious, conservative ruling classes resisted the Radical Right prescription, they were trapped in the logic of their own decision to assign such a high priority to territorial revisionism that association with Hitler to achieve it was deemed acceptable. Their hope of exploiting German power to restore Hungary's historic frontiers while avoiding identification in Allied eyes as Germany's partner was quite unrealistic: it overestimated their own dexterity, underestimated German alertness, and trivialized the wartime seriousness of the Anglo-American-Soviet alliance. It also lacked integrity. The Germans readily capitalized on this flaw in Hungarian policy by letting it be known that their partition of Transylvania in August 1940 between Hungary and Romania was but provisional and that they would be inclined to award that entire province at war's end to the satellite partner that made the bigger contribution to the Axis war effort. This led to the bizarre and tragic result that Hungary contributed an army corps to Hitler's invasion of the Soviet Union in 1941—though it had no tangible war aims there—lest it be outclassed by Romania in the competition for Hitler's favor over Transylvania. Toward the end of that year, Britain gave Hungary an ultimatum to withdraw from the Russian campaign and declared war when it was ignored; a few days later, Hungary declared war on the United States, which did not reciprocate for six months.

Yet, until 1944, Hungary's war—especially against the Western Allies—was rather formal and stylized. At the beginning of 1943, the Hungarian corps in Russia was decimated by the Soviet army at Voronezh and then virtually abandoned by its *Wehrmacht* ally during the retreat from Stalingrad. This provided the Hungarian government with a pretext to withdraw the remnants into Hungary by April 1943, after which date only a few rear-area garrisons remained in the Soviet Union and the bulk of the Hungarian army was manning the Carpathian passes against the ostensible Romanian ally as well as the Soviet foe. Just as Hungary and Romania had competed for Hitler's favor when the Axis tide was running strong, so after that tide turned to ebb, they competed by shirking their obligations to him, with each rationalizing the thinning of its military contribution on the eastern front by arguing that its security against the other required the concentration of its army at home.

As for Hungary's formal state of war against the British and Americans, their planes were not fired on when overflying Hungary and, in turn, did not bomb it until 1944. Before that year, indeed, Hungary was virtually a neutral in the war between the Axis and the Western Allies. And in such noncombatant war efforts as industrial production, the export of food and raw materials to Germany, and the facilitation of the *Wehrmacht*'s communications and supply systems, Hungary minimized its contribution to the Axis effort. Only one-fifth of its electric power was allocated to war production, and only one-quarter of the stipulated food delivery quotas had been dispatched to Germany before 1944.

To the chagrin and rage of the Radical Rightists, domestic social and institutional coordination with the Nazi model was also diluted by the ruling conservatives. Parliamentary debate was vigorous, opposition parties were active, trade unions remained free, the press was lively—though overt criticism of Germany was taboo. Civil liberties endured. Escaping Poles and Allied war prisoners received shelter, and the Jews, though economically and socially molested, were shielded from extermination. Finally, the exasperated Hitler occupied Hungary in mid-March 1944 and forced the replacement of the foot-dragging and peace-seeking conservative government with a more pro-German one, though still not with an all-out Radical Right one. Until then (but not thereafter), the domestic bearing of his Hungarian satellite had been in relatively decent contrast to the gadarene behavior of its Croatian, Romanian, and Slovak neighbors, which were also Nazi Germany's satellites.

Now finally began a phase of emphatic Hungarian coordination with the German war effort and with Nazi policies generally. Trade unions as well as democratic and Socialist parties were at last dissolved. (The Communists had all along been illegal and underground.) Anti-German political and cultural leaders were arrested and occasionally killed. The Jews of the provinces (though not those of Budapest, who were protected by Regent Miklós Horthy, the Swedish businessman-diplomat Raoul Wallenberg, and some ecclesiastics) were deported to Auschwitz for extermination. The economy was seriously mobilized for war, and armies were sent to the rapidly approaching eastern front. All this was the work of the new Hungarian authorities themselves, not simply of their German advisers.

There was a logic—albeit a fatal one—to this belated Hungarian acceptance of synchronization with Nazi Germany. The failure of the

Western Allies to protect Italy from Nazi German invasion and vengeance at the time of its abortive attempt to quit the war in September 1943 had given a sobering lesson to its would-be Hungarian emulators. And by December 1943, Beneš had persuaded the Big Three to restore Czechoslovakia's prewar frontiers essentially intact, thus depriving the Hungarians of another potential incentive to desert Hitler—permission to retain their territorial gains of 1938 and 1939. Similarly, the Allies endorsed Yugoslav intentions (articulated by the royal government-in-exile as well as by Tito's Partisan movement inside that country) to undo the Hungarian reannexation of part of their land in 1941. Only in regard to the Transylvanian issue, where Hungary's rival was its fellow Axis satellite Romania, were the Allies still uncommitted by 1944. Hence the seeming plausibility of the position taken by the government that Hitler imposed on Hungary in March 1944, the position that there was simply no feasible alternative to the partnership with Nazi Germany. And there is no denying that the masses implemented this logic with vigor: the workers henceforth worked hard, the soldiers fought hard, and the general population applauded the extermination of the Jews. The Western Allies, in turn, for the first time treated Hungary as a full-fledged enemy state and bombed its cities heavily.

Romania's leap out of the Axis and into the Allied camp, executed on August 23 to 25, 1944, dealt a damaging blow to Hungary in both military and political dimensions. By opening the Balkan Peninsula to the Soviet armies, it facilitated and hastened their invasion of Hungary. And by executing this volte-face so efficiently and decisively, the Romanians enhanced their claim, in Allied eyes, for the eventual reversion of all of Transylvania. In response, Regent Horthy of Hungary sought to dismiss the pro-German ministry that Hitler had imposed on him half a year earlier and to conclude an armistice with the Allies. But his preparations for this coup were so inadequate and clumsy that the Germans easily thwarted it and finally installed a thoroughly reliable and enthusiastic, albeit wild and psychopathic, Radical Right regime. The main body of the Hungarian army remained loyal to this pro-Nazi regime and, in partnership with the *Wehrmacht*, resisted the Soviet armies tenaciously until the war's end. The siege of Budapest during the winter of 1944/45 was a fiercely fought encounter. Not until April 4, 1945—a month before the end of the war in Europe—were the German and Hungarian forces pushed off their last toehold in Hungary and into Austria. Meanwhile, in the part of Hungary already conquered by the Soviet army, a provisional government had been unveiled on December

23, 1944, composed of three maverick generals who had heeded Horthy's abortive armistice instructions as well as representatives of the Communist, Social Democratic, Smallholder, and National Peasant parties. This coalition styled itself the National Independence Front and committed itself to a relatively moderate program combining land reform (the breakup of Hungary's traditional system of large estates) and universal, equal, direct, and secret suffrage (also an innovation in Hungary) with protection of private property and private economic initiative. It concluded an armistice with the Allies on January 20, 1945, and transferred itself from provincial Debrecen to the freshly liberated capital of Budapest on April 14.

5

World War II engulfed Yugoslavia in the spring of 1941 despite—or, perhaps, because of—the desperate diplomatic gymnastics of its government to keep it at bay. By that time, France had fallen, in June 1940, and Yugoslavia's neighbors, Hungary, Romania, and Bulgaria, had joined the Axis' Tripartite Pact respectively on November 20 and 23, 1940, and March 1, 1941. By that time also, Mussolini's attack on Greece, launched via Albania on October 28, 1940, had been repulsed, and Hitler felt obliged to come to the rescue of his foundering partner. Hence he applied strong pressure on Yugoslavia, which was already surrounded by Axis states and trapped in Germany's economic web, to adhere to the Tripartite Pact in order to secure his diplomatic rear in the Balkans preparatory to his military assault on Greece and the British contingents bolstering Greek resistance.

To purchase Yugoslavia's adhesion, Hitler was even prepared to guarantee its territorial integrity and to give assurances that it would not be required to furnish military assistance or even to let its territory be used for the stationing or transportation of alien troops. Compared with that of its neighbors, Yugoslavia's status within the Axis camp would thus be uniquely privileged. Other, more ambiguous, bribes were also held out by Hitler: a possible Yugoslav territorial corridor to the Aegean Sea, with the eventual transfer to Yugoslavia of the Greek port of Salonika; and, on a more personal level, a suggestion of German support for the regent, Prince Paul, should he wish to postpone the young King Peter's legal coming-of-age from his eighteenth to his twenty-fifth birthday or otherwise tamper with the royal succession. Simultaneously,

Belgrade was given to understand that it would no longer be permitted the luxury of categorical neutrality. Unless the invitation to join the Tripartite Pact were accepted, Yugoslavia would henceforth be regarded by the Axis powers as an unfriendly state. This meant, at a minimum, that Hitler would authorize his Italian, Bulgarian, and possibly Hungarian allies to present and enforce their irredentist demands to several Yugoslav regions.

Whether or not he was tempted, Prince Paul remained aloof to the personal bait. But the Germans' political arguments and the power behind them impressed him and the cabinet. To repudiate the Axis pressure would have required a degree of domestic consensus and national solidarity that the regime could not command and that it was reluctant to even try to mobilize, lest it be swept aside in the ensuing momentum. In international terms, furthermore, Yugoslavia was geopolitically surrounded by the Axis and militarily unprepared to resist the *Wehrmacht*. Quite apart from profound errors in the strategic dispositions of its armed forces, which would be exposed only in the course of the April campaign, their logistical situation was known to be hopeless ever since Hitler's destruction of Czechoslovakia had interrupted Yugoslav access to the production of the Škoda works, the Yugoslav army's standard supplier of armaments. No anti-Axis power was able to fill this gap, and Yugoslavia's own industrial capacity and transportation network were inadequate to sustain mechanized warfare. The northern frontiers and cities would probably have been indefensible even without the additional aggravation of *Wehrmacht* concentrations in Romania and Bulgaria. Nevertheless, due to Croation and Slovene political sensitivities and pressures, the Yugoslav armies were unable to concentrate in the mountainous central redoubt of Bosnia and western Serbia. That would have been the logical strategy, given their technical limitations. Instead, they were deployed in shallow and overextended formations all along the long borders.

Aware of Yugoslavia's military weakness and economic vulnerability, and of the current European balance or absence of it (France defeated, Britain unable to render effective aid, the United States remote, and the Soviet Union enigmatic), the government cautiously eased its way toward the formidable Reich and eventually agreed to the Tripartite Pact on March 25, 1941, under the exceptional waivers promised earlier. Though many Western and virtually all Communist historians have deplored this decision as short-sighted—which, in retrospect, it was—and dishonorable, it should be recalled that unlike Neville Chamberlain and Edouard Daladier in 1938 and Stalin in 1939, Prince Paul in 1941 was negotiating with Hitler only over his own country's destiny and not bartering away another's territory or existence and that German power over Europe had meanwhile vastly expanded. The British were prodigal with exhortations to the Yugoslavs to preserve their honor by rebuffing the Germans, but short on the military assistance that might have given weight to this advice. Simultaneously, Britain was showing greater patience with Sweden, which permitted the movement of German troops and material across its territory to and from occupied Norway, than with Yugoslavia, which had extracted from Germany explicit exemption from such collaboration. It is scarcely surprising that Prince Paul, for all his Anglophilia, came to suspect that London regarded Yugoslavia as expendable.

Had he been leading a united country that was persuaded of the utility of his diplomatic and domestic policies, Prince Paul would have been able to shrug off the foreign excoriators of his accommodation to Germany. Alas for him, the Serbs—ranging from the Orthodox clergy, the secular intelligentsia, the Sokol youth, the veteran organizations, the officer corps, to the masses in general—were already disgruntled by his earliest concessions to the Croats and now doubly resented his adhesion to the Tripartite Pact as a betrayal of their national interest, tradition, and honor. What outraged them was not so much the specific terms of the pact—Hitler's waivers, after all, meant that Yugoslavia was giving little away, while his record of broken guarantees also suggested that it was gaining little in the long run—but its deeper significance as capitulation to Teutonic pressure, a capitulation that they abhorred as cutting against the grain of Serbia's finest history. The ink was scarcely dry on Yugoslavia's signature when a bloodless military coup overthrew the regime during the night of March 26/27, 1941, forced "the foreigner" Prince Paul into exile, proclaimed King Peter II as having come of age, and installed General Dušan Simović as premier. Simović had recently served as chief of the general staff (1938–40) in between two terms as commander of the air force (1936–38, 1940–41) and had warned Paul as recently as March 23 that adherence to the Tripartite Pact might well provoke the air force mutiny that had now occurred.

While the Serbian crowds greeted the coup ecstatically, the Slovenes and Croats appeared to be resentful that the Serbs had unilaterally committed the whole country to the very high—virtually certain—risk of war. Whereas moderates profoundly regretted this development, more extreme Croatian nationalists by now actually craved a German invasion as a catalyst to achieving Croatian independence. And, indeed, the extremists would presently serve the Axis as auxiliaries in the demolition of the Yugoslav state.

Simović desperately sought to stave off German retribution. Despite the anti-Axis enthusiasm of the Serbian street crowds and the embarrassing applause with which Churchill greeted the coup, Simović frantically tried to make his government and its policies palatable to the Axis powers. His foreign minister promptly announced that the postcoup government would honor all international obligations contracted by its predecessors, while Simović himself sought to convince the German minister to Belgrade that the coup's causes were exclusively internal—that is, pertained to the country's endemic Serbian–Croatian frictions—and that he was a friend of Germany and proud of his acquaintance with his fellow aviator, Reichsmarshall Hermann Goering.

As far as Hitler was concerned, all these exertions by the Yugoslav leaders to appease him were futile. Within hours of the coup, he directed the *Wehrmacht* to smash Yugoslavia, without regard to any possible declarations of loyalty that its new government might tender. He simply ignored the new Yugoslav foreign minister's abject pleas to be permitted to come to Berlin for consultations and was undeterred by a last-minute demonstration of Stalin's interest in the form of the Soviet-Yugoslav Treaty of Friendship and Non-Aggression, concluded late on April 5. The next morning, the *Luftwaffe* razed Belgrade, and the German army poured across the Yugoslav borders from sundry directions. The Croatian troops widely shirked mobilization, mutinied sporadically, and scarcely fought, while the Serbs' performance was flabby in comparison which their tenacity in World War I. Yugoslavia's ultimate defeat was admittedly a foregone conclusion, yet the feeble performance of its 2-million-man army in 1941 was a sorry reflection of its people's current sentiments toward the regime and its system. A decade of repressive yet irresolute misgovernment, in the name of national unity, authority, and regeneration, had squandered the confidence of all ethnic communities, even the Serbs, while belated concessions had not purchased even the elementary loyalty of the Croats. On April 10, an independent Croatian state was proclaimed in Zagreb; on April 12, Belgrade fell; and on April 17, the remains of the Yugoslav army surrendered, and the government went into exile (first in Athens, then in Jerusalem, and finally in London).[9] Outraged by the Belgrade coup, Hitler had taken swift and devastating revenge on those who had thus dared to cross him. Yet in the long run, he may have been the biggest loser of the Balkan campaign of 1941, since it postponed his scheduled invasion of the Soviet Union by a month. Though it is not possible to confirm or refute definitively the following proposition, the loss of that precious month of summer campaigning may have proved decisive when his panzers ground to a halt at the outskirts of Moscow that winter. Hitler might have been shrewder had he explored the Simović government's conciliatory gestures.

Judging the Yugoslav events and decisions of March 1941 remains as difficult and controversial a problem for historians as it was at the time for the participants. Some argue that in reluctantly agreeing to the Tripartite Pact, Prince Paul's government was using diplomacy as the necessary substitute for the economic and military power, as well as the national unity and international assistance, that Yugoslavia lacked at the time; that the exceptional waivers it extracted from Hitler were likely to avert the war from Yugoslavia's borders and hence to preserve the state from annihilation and its peoples from devastation; that this government's policy, in short, was rational and prudent. They proceed then to denounce the coup for being exclusively Serbian and for precipitating the Axis invasion and occupation, which, in turn, led to a devastating cycle of guerrilla resistance and reprisals, ethnic and ideological civil war, reciprocal atrocities and exterminations, and ended in Communist rule of the country. In the light of these consequences, they argue, the coup must be condemned as politically immoral and counterproductive.

Simović's defenders retort that even in politics, morality is not simply a matter of weighing material gains and risks and that the coup restored Yugoslavia's self-respect and Serbia's honor after the allegedly shameful abdication of signing the pact. (In a sense, this debate replicates the earlier one comparing the ethics and pragmatism of Polish defiance with Czech accommodationism in World War II.) They point to Romania as an example of a country that had "realistically" tacked with every wind of the 1930s and 1940s, yet also ended as a battleground, though admittedly it did not suffer the devastation of Yugoslavia's civil wars. They remind their rivals of Hitler's record of broken commitments, of the unpopularity of the Pauline regime, and of the genuine revulsion of the Serbian masses for the pact. They claim that despite its *Putsch* origins and its gerontocratic composition, the short-lived Simović government was interwar Yugoslavia's most authentically representative one. Though uncomfortably aware that Simović's coup was scarcely applauded outside Serbia, they insist that since the Slovenes and Croats had even better reason to fear Italy than the Serbs had to distrust Germany, its rebuff to the Axis expressed the interests of all three Yugoslav state-peoples. Here they accept Hitler's and Churchill's parallel interpretations of the coup as a reversal of the previous government's foreign policy and ignore, or interpret as merely tactical, Simović's failure to repudiate the Tripartite Pact and his abortive endeavor to propitiate the Axis dictators.

To the criticism that its denouement was but a Communist dictatorship, the coup's non-Communist defenders answer that the Balkans would in any event have fallen within the Soviet sphere at the conclusion of World War II and that Yugoslavia's own Tito was, after all, preferable to a Stalinist agent imposed by the Soviet Union. To this, a disinterested observer might add that it was precisely the conditions of guerrilla warfare, which were so deplored by the defenders of Prince Paul's policy, that stimulated a Communist Yugoslav patriotism, a sense of self-reliance, and an effective administrative structure among Tito's Partisans and hence provided him with the organizational, psychological, and emotional resources to defy Stalin in 1948. Communist contributors to the discussion make some of the same points as the coup's "bourgeois" defenders, but in general they deemphasize the autonomous role of the *Putsch* officers and insist that they could have acted as they did only thanks to the revolutionary rage of the masses, incited by the Communist party, against the Pauline regime's policy of commitment to the Tripartite Pact.

But we have allowed the historical debate to run slightly ahead of our narrative, for the phenomenon of Tito—let alone the rift between Tito and Stalin—was quite unforeseen in the spring of 1941.

Though Hitler's conquest of Yugoslavia was extraordinarily swift and easy—requiring just slightly more than one week—and though his vastly more prodigious and portentous invasion of the Soviet Union would soon relegate this vanquished Balkan country to a strategic backwater, it was to be rendered a peculiarly expensive backwater by the resistance movements that soon emerged and that by 1944 tied down 125,000 German troops.[10] Most of the resistance movements arose in defense of one or another of the ethnohistorical territories out of which the Yugoslav state had been amalgamated in 1919. But one of them, the Communist Partisan guerrilla movement led by Josip Broz Tito, astutely committed itself to an all-Yugoslav strategy, even though the ethnonational tensions that had wracked this state throughout the interwar decades seemed to render such a strategy rather unpromising. This decision put the Partisan movement on a collision course with the more provincial and parochial defense movements and with Hitler's overall disposition of his Yugoslav conquest.

Proceeding roughly from northwest to southeast, those dispositions by Hitler of the several fragments of defeated Yugoslavia were as follows: (1) Slovenia was partitioned between the German Reich (which annexed its northernmost share into the Austrian districts of Styria and Carinthia) and Italy. Though the territorial proportions were approximately in the ratio of 2:1 in favor of Germany, the Italian share did contain the Slovenian capital city of Ljubljana. Furthermore, while the Germans would pursue a policy of demographic and linguistic Germanization in their annexed area, Italian policy tended to be relatively supportive of Slovenian ethnoculture in its zone. (2) Hungary reannexed a substantial part of those territories that had historically belonged to it before 1919, and in this area it vigorously resumed the old policy of seeking to denationalize the local non-Magyar populations—a policy culminating in a massacre of Serbs and Jews in January 1942 that had the unintended effect of swelling support among the survivors for Tito's Partisans. (3) Another part of the erstwhile Hungarian lands of interwar Yugoslavia, the Banat (to which both Hungary and Romania now aspired), was placed under an autonomous provisional German military administration, for which the local *Volksdeutsche* furnished the auxiliary personnel. (4) Italy annexed most of the Adriatic islands as well as the Dalmatian littoral, while (5) the rest of (mainly inland) Croatia was established as an ostensibly independent puppet state of the Axis. Though Hitler allocated Bosnia-Hercegovina to this Croatian satellite, the Croatian public resented the assignment of the Dalmatian coastal strip and the adjacent islands to Italy. (6) Montenegro was given autonomy under Italian supervision (not annexation) and (7) western Macedonia and the Kosovo region of southwestern Serbia were annexed to Italian-occupied Albania. The local population there was, indeed, largely Albanian. (8) The rest, bulk, of Macedonia was assigned to the Axis's Bulgarian satellite. (9) Finally, rump Serbia, the political core of interwar Yugoslavia, was established as a puppet state headed by a collaborationist general named Milan Nedić, whose intended function was roughly analogous to that of President Emil Hácha in the Protectorate of Bohemia and Moravia. German or Italian garrisons were stationed in all these fragments of Yugoslavia, no matter what the degree of each one's formal autonomy.

These several dispositions of the Yugoslav lands not only were complicated, but also proved to be highly controversial within the Axis camp. Indeed, for the rest of the war, persistent frictions over the Yugoslav spoils were to poison relations among the several Axis partners and their local protégés—and thus to supply political grist for Tito's mills. An early contribution to these internecine frictions was the German decision to free all the non-Serbians among their Yugoslav prisoners of war.

The German decisions (1) to permit the hitherto rather peripheral Ustaša (Insurgent) anti-Yugoslav secessionist movement to establish itself as the government of the new, formally independent, puppet state of Croatia, and (2) to assign the ethnically mixed region of Bosnia-Hercegovina to this Croatian state and regime soon emerged as a major political blunder. Exhibiting the rigid extremism and myopic xenophobia that is often characteristic of nationalistic politicians who have languished in exile from their native constituency (in this case as protégés of Mussolini in Italy), the Ustaša leadership promptly proceeded to alienate the majoritarian Croatian peasant movement, to exterminate the Jews, and to persecute the substantial Serbian Orthodox population of Croatia proper and of Bosnia-Hercegovina, thereby driving these Serbs into the ranks of the Partisan resistance movement. It also compromised itself with the nationalistic Croatian masses (who otherwise applauded Croatia's gaining of independent statehood) by ceding the Dalmatian littoral and Adriatic islands to Italy and by offering to a member of the Italian royal house the formal kingship of Croatia. (This nobleman, Duke Aimone of Spoleto, upon whom the Ustasa conferred the title of King Tomislav II of Croatia, had the prudence never to set foot in his putative realm.) Further discrediting the Ustasa regime in the eyes of its own potential constituency were the privileges that it accorded to the local *Volksdeutsche* minority, its requisitioning of the 1942 harvest for shipment to Germany, and its humiliating inability to dislodge Tito's Partisans from Bosnia-Hercegovina. Finally, the imposing Soviet military successes after 1942 reawakened a Croatian Slavophilism that also undermined the Ustaša regime, with its all-too-vivid dependence on Germany and Italy. Indeed, the German military authorities repeatedly warned Hitler that the Ustaša's combination of narrowness, incompetence, savagery, and dependence was undermining rational German interests—to no avail. Hitler backed the Radical Rightists despite their dysfunctional impact on his war effort earlier than he did elsewhere among his satellites.

Indeed, a peculiar reversal of the interwar pattern of German and Italian stances toward Yugoslav issues now emerged, with Croatia as its fulcrum. Before 1941, Mussolini had generally sought the breakup of Yugoslavia and had often subsidized the Ustaša (and other secessionists) as a lever with which to subvert it, whereas Hitler had deemed German economic and political interests to be best served by bringing the whole of an integrated Yugoslavia into his orbit and had therefore backed the Serbian centralizers. After the 1941 partition of the country the Italian authorities, both in Rome and locally, noted an aggrieved Croatian irredentism directed against their own recent annexation of the Dalmation coast and accordingly switched to protect and support the Serbian victims of and resisters against the Ustaša terrorism. Hitler, however, still enraged by the Serbian defiance of his will in March and April 1941, now took up his old "Austrian" heritage of endorsing the Croats' contempt for the allegedly primitive and inferior Serbs, even though this ideological endorsement undermined his objective interest in pacifying the Balkans for the duration of the war. The *Wehrmacht* officers on the spot did not find it easy to accommodate themselves to Hitler's perspective.

The moral and psychological impact of the April defeat was even more devastating for the Serbs than for the other Yugoslav ethnonations. Their renowned army had been easily rolled up; their seemingly powerful state apparatus, blown away. Out of this initial bewilderment there eventually grew a generalized disillusionment with the entire interwar system and its ruling regime. This readiness for new political alternatives was even more palpable among the Serbs outside Serbia—those of Croatia, Bosnia-Hercegovina, Montenegro, and the Hungarian-occupied zone—than those of rump Serbia proper. It was, of course, strengthened as the war progressed by exemplary Soviet Russian feats of arms, and its eventual beneficiary would be the Communist-led Partisans. But before they could succeed in winning Serbian as well as non-Serbian endorsement of their political program and their bid to rule a new Yugoslavia, the Partisans had to bide their time while other, ethnically more parochial, elements tried their hand at rousing the Serbs out of their initial postdefeat depression.

The *Wehrmacht* had been in such haste in April 1941 to push on to Greece and Crete, and thereafter to redeploy for the invasion of the Soviet Union, that it had failed to disarm effectively all tactical units of the smashed Yugoslav army, some of which took to the hills and woods as so-called Cetniks—a traditional Serbian term stemming from Ottoman times and denoting a combination of guerrilla band and village militia. Many of these Četnik units soon put themselves at the disposal of General Nedić's collaborationist regime, hoping thereby to enable him to function more effectively as a buffer between the German occupation authorities and the Serbian population. Others were for a while more resolute in initially resisting the Germans. Of these, the most important was commanded by Colonel Draža Mihajlović, whose early guerrilla exploits seemed to be sufficiently vivid to earn him appointment as war minister in the royal government-in-exile in London in January 1942, though physically he remained in the Serbian woods and Montenegran hills. But Mihajlović's main energies were soon devoted less to resisting the German occupation than to combating the Croatian secessionists and protecting Serbs from Ustaša atrocities, and this not in the name of a new, authentically transethnic, Yugoslavism, but in the discredited image of interwar Pan-Serb hegemony. That the exile government was hopelessly divided by Serb-versus-Croat issues compounded his difficulties.

In September and October 1941, Mihajlović held several inconclusive conferences with Tito. Their outcome symbolized the incomprehension and confrontation between the old and the emergent Yugoslavias.

Tito's mettle and strategy had already been indicated four years earlier when, on becoming secretary-general of the Yugoslav Communist party, he elected to bring its Central Committee home to Yugoslavia from Moscow and Vienna, to share the rank-and-file's underground tribulations, risks, and opportunities. The Axis invasion had found him in the Croatian city of Zagreb, from where he made his way to Belgrade a month later. In September, he left the capital and joined his Partisan guerrillas (who had staged their first attack on a provincial German outpost in July) in the wooded hills of western Serbia. (There are analogies here to Mao Tse-tung's earlier abandonment of China's urban centers for its rural periphery, with its more promising potential for guerrilla warfare and social revolution.)

Whereas Tito's Partisans were volunteers and hence highly mobile, Mihajlović tended to recruit the entire male population of villages into his Četnik units, which reduced their mobility and rendered them far more vulnerable to German reprisals. Accordingly, when the Germans massacred several thousand Serbs in the provincial town of Kragujevac on October 21, 1941, in reprisal for a Partisan ambush elsewhere, Mihajlović lost his nerve for further harassment of the *Wehrmacht*, lest it lead to the decimation of the Serbian nation. He decided that henceforth his responsibilities were to protect Serbs from Ustaša atrocities, to collect arms, and to husband his resources for the day when the royal government-in-exile would be enabled to return home by the strategic defeat of Germany by the Allied Great Powers. Since that Allied victory would eventually compel the Axis armies to evacuate Yugoslavia, he found little difficulty in slipping, in time, into an interim *modus vivendi* with the local *Wehrmacht* and Italian commanders (and with their protégé Nedić, his fellow Serb officer) and in treating as his major foe the Communist Partisans, who, after all, were a native threat to most of his political and social values. He was confident that his status as war minister in a government supposedly close to the Western Allies would screen this incipiently collaborationist stance.

Tito's strategy was both militarily and politically innovative, albeit equally calculating. He accepted, indeed welcomed, the fact that his constant harassment of the German forces provoked them to indiscriminate mass reprisals, from which the villagers could escape only by joining his mobile Partisan columns. Pushed out of western Serbia at the end of 1941, the Partisans migrated first to Montenegro and then to Bosnia, where they transcended the fierce local ethnonational enmities by protecting all from all—Serbs from Ustaša atrocities, Croats and Muslims from Četnik vengeance—and molding volunteers from all the communities into a common Yugoslav phalanx against the Axis occupiers. Although several German, Italian, Ustaša, Četnik, and Bulgarian punitive offensives kept Tito's forces ever on the move, liberated zones were carved out in which Communist-controlled but authentically participatory political mechanisms and popular institutions were developed. That these early liberated Partisan zones were within the nominally sovereign state of Croatia (that is, in Bosnia-Hercegovina) further discredited its puppet Ustaša regime. Indeed, by the spring of 1943, the Ustaša controlled little outside the city of Zagreb, as many younger Croats responded to Tito's astute cultivation of the revolutionary tradition of the brothers Ante and Stjepan Radić, the founders of the populist Croatian peasant movement. After the fall of Mussolini and the surrender of Italy in the summer of 1943, the liberated zones were extended to Montenegro, Dalmatia and the islands, and, intermittently, southern Slovenia. Only in rump Serbia proper did Tito fail to establish himself until after the Romanian volte-face of August 23 to 25, 1944, forced the *Wehrmacht* to redeploy its Balkan garrisons.

On November 26 and 27, 1942, Tito formalized his political challenge to the Axis partition of Yugoslavia, to the several occupation regimes, as well as incipiently to the royal government-in-exile by establishing the Antifascist Council of National Liberation of Yugoslavia (AVNOJ) in the liberated northwestern Bosnian town of Bihać.

Composed of Communist and non-Communist delegates, AVNOJ declared itself to be the legitimate administration in the liberated zones, committed itself to a federalist solution for the country's ethnonational problems—thus making a constructive appeal to the regional and historical identities that other actors were exploiting in divisive directions—and promised a democratic postwar regime that would protect private property. Local and provincial national liberation committees were authorized throughout the liberated zones. A year later—having survived several fierce Axis offensives in which Mihajlović and Nedić blatantly collaborated with the *Wehrmacht* and the Ustaša in desperate efforts to destroy their common Partisan enemy, and having thereby earned British recognition and support as the most effective and reliable Yugoslav resister to Hitler—Tito escalated his political claims by having AVNOJ declare itself to be the Provisional Government of all Yugoslavia (with himself as premier, defense minister, and marshal), deposing the government-in-exile, and prohibiting King Peter from returning to the country pending a plebiscite. These political decisions, taken at Jajce in central Bosnia on November 29, 1943—at the very moment when the first conference of the Big Three at Teheran resolved to support the Partisans' military endeavors—earned Tito a sharp rebuke from Stalin, who feared that they would unnecessarily ruffle feathers in London and Washington. A year earlier, Stalin had vetoed Tito's wish to take similar political steps at Bihać, so this time Tito did not inform Moscow of his intentions. In the event, Stalin's concern proved groundless, as the Western Big Two did not protest the Jajce decisions. Nevertheless, the effrontery of Tito's independence appears to have rankled in Stalin's mind. And in Tito's mind, there lingered a reciprocal awareness that his Partisans had received their earliest and most substantial military help from the British rather than the Soviets.

Though heavy fighting persisted through 1944—indeed, the Germans, emulating Tito's own mobile tactics, came within a hair of wiping out his headquarters in May—on the political level, that year marked a sustained series of victories for the Partisan movement, as the Yugoslav peoples rallied massively to Tito's perceptive combination of patriotic, revolutionary, and federalist slogans and programs. Though the hapless Mihajlović continued to help evacuate British and American airmen forced to bail out over Četnik areas, his authority disintegrated as his men deserted either to the more actively resisting Partisans or to explicitly fascist collaborators, or degenerated into sheer brigandage. On May 17, under British pressure, the government-in-exile dropped him as its war minister, and on September 12, King Peter broadcast from London an appeal to his subjects to rally to the Partisans for the final expulsion of the Axis. Also in September 1944, many members of the Croatian Home Guard militia deserted to Tito in response to a limited amnesty promise-cum-ultimatum that he had issued to them. Since at this stage of the war the undisciplined Četnik bands were massacring Croats and Muslims in revenge for the earlier Ustaša atrocities against Serbs, the Partisans again earned political capital by their sagacious and shrewd policy of protecting the victims of such interethnic violence and thus projecting themselves as the bearers of the brotherhood of the Yugoslav peoples.

In that same year of 1944, the Partisans definitively established their control in the peripheries of Yugoslavia—the Hungarian-occupied zone, Macedonia, and Slovenia. In Macedonia, interestingly enough, the traditional rivalry between Serbia and Bulgaria for dominion was reflected within the corresponding Communist parties, as the Bulgarian Communists sought to utilize Bulgaria's wartime annexation of most of Macedonia to contest the legitimacy of Tito's Partisans operating there. To no avail. Yugoslav Macedonia was needed and utilized by Tito as a showcase for his federalist reconstruction of Yugoslavia in place of the discredited former Serbian centralism. In Slovenia, too, federalism was a powerful Titoist asset, one that he was even able to double when Hitler exploited the fall of Italy to incorporate Venezia Giulia (with Istria and Trieste) into the Reich (shades of the Holy Roman and Habsburg empires!). Tito's claims to these lands and to parts of Austrian Carinthia then established his moral authority with Yugoslavia's Slovene ethnonationalists.

Although the preceding pages have stressed Tito's successful and authentic projection of his roles as anti-German liberator and transethnic reconciler, it would be a mistake to overlook the third leg of his politico-ideological triad, the Communist revolutionary one. Already in the summer of 1941, at the moment of his launching his Communist party's Partisan movement, he had stated that its purpose was "to prepare to seize power, and to seize it in such a way that the bourgeoisie would never regain it."[11] From this goal he never deviated, despite Stalin's concern that its very blatancy would jeopardize the global as well as the local anti-Nazi coalition.

One of the structural reasons why this revolutionary goal was acceptable to many Yugoslavs who were otherwise not inclined to Communism has already been mentioned—the ignominious collapse of the interwar regime and army in April 1941, which seemed to expose the interwar sociopolitical system as a kind of bluff, sham, or fraud. Sociological reasons why the Partisans could win authentic support without denying their Communist aspirations were the appeal of collectivism to Yugoslavia's mountaineer folk, the attraction of the image of wealth through industrialization to marginalized peasantries, and the readiness of hitherto apathetic, alienated, or "prepolitical" sectors of the population to be politicized in the turbulent, yet exhilarating, setting of war. Contingent organizational factors were the Communists' paramilitary and administrative experience gained during the Spanish Civil War, their discipline, ideological commitment, and conspiratorial skills, and the personal quality of Tito's leadership. Finally, all their native competitors having compromised themselves by collaboration with the occupiers and/or by participation in interethnic atrocities, only the Partisans could plausibly project themselves as embodying a transethnic, all-Yugoslav patriotism.

The historical literature on World War II often draws analogies and contrasts between Yugoslavia and the other states of East Central Europe, especially between Yugoslavia and Poland. But this should be done with prudence as well as imagination. For example, the frequent analogy between Mihajlović's Četniks and the Polish Home Army is flawed. Admittedly, both were ultimately abandoned by the Western Allies, who had originally endorsed them; but the Polish Home Army was by and large untainted by collaboration, and hence its retrospective moral grievances are stronger than are the Četniks'. A more valid analogy may be drawn between the fates of the two governments-in-exile in London. Under Western Allied pressure, certain members of each were permitted to return home and join the Communist-controlled governments that had established themselves on the ground. But these individuals were soon neutralized—even sooner in Yugoslavia than in Poland—and never exercised decisive political leverage, though the symbolic effect of their "sanitizing" the regimes in the eyes of Western and local public opinion was, for an interval, quite useful to the Communists. A more vivid and tragic analogy between wartime Poland and Yugoslavia lies in the degree of their suffering. Some telling data for Poland were presented earlier. In Yugoslavia, the war—which was an interethnic civil war and an ideological social revolutionary war as well as a war of national liberation—was also hideously destructive. Approximately one and three-fourths of a million people—over a tenth of the population—were killed, half the livestock and the railway trackage and stock, three-fourths of the plows and the railway bridges, over three-fifths of the road surfaces, a third of the industrial value, and a fifth of the housing stock were destroyed. And yet, out of all this carnage and violence, this hatred and disunion, there emerged endurance, heroism, and a new national unity and sense of purpose.

Against this background, we now finally examine a claim that is frequently made by and on behalf of the Yugoslav Partisans: that they liberated their country from Axis occupation by their own efforts, that—in contrast to the postwar Communist regimes in other states of East Central Europe—their administration was not imported and installed by the Soviet armies, that therefore it earned greater legitimacy in the judgment of its own public than did the Soviet-installed Communist regimes, and that it was this authentic legitimacy deriving from the Partisans' performance as the wartime self-liberators of Yugoslavia that enabled them to defy Stalin in the late 1940s. While it is, of course, true that the wartime exploits of the Partisans were indeed heroic and that during the fighting they developed the politico-administrative apparatus and élan that enabled them to gather and consolidate their power, it is surely an extravagance to claim that they evicted the *Wehrmacht* from Yugoslavia in any overall strategic sense. It was the advances of the Soviet armies through Romania and Hungary in the autumn of 1944 that obliged the *Wehrmacht* to evacuate the Balkans, lest its forces there be cut off from the Reich. By then the Partisans, who had indeed harassed, tied down, and bloodied but not defeated the Germans, were strong and confident enough to take over the country without foreign sponsorship or intervention.

6

The outbreak of World War II found Romania governed by a royal dictatorship, steered by King Carol II, which competed for public support with a genuinely popular Radical Right movement that was conventionally termed the Iron Guard. The propaganda myths of the royal dictatorship were organic nationalism, family, church, and the gospel of work. Its rubber-stamp parliament, for example, was selected from the three work categories of agriculture, industry and commerce, and the professions of the intelligentsia. All in all, it was a pseudo-radical, semifascist burlesque, intended to stymie the Iron Guard and steal its ideological appeal. It benefited economically from the competition, as war approached, among Britain, France, Germany, and Italy for Romania's foreign trade, especially its oil exports. While King Carol and a few of his cronies exploited this favorable situation to amass huge personal fortunes, prosperity also became somewhat more general, as the oil-export revenues were applied to rapid, state-directed industrialization.

Diplomatically, the royal dictatorship—which had been inaugurated on February 10, 1938, on the eve of Hitler's Anschluss of Austria—adopted a posture of straddling, as Romania edged away from its hitherto categorical identification with a now weakening France and sought to balance between the Western and the Axis powers. This quest entailed the delicate task of affirming Romania's own substantial territorial acquisitions under the Paris treaties system that had closed World War I, but without irritating a generally revisionist Germany, and it came closer to success than is often appreciated. Enjoying many ready markets for its oil, Romania could stave off German economic hegemony longer than its neighbors could, while its generous treatment of its German minority, relative to Hungary's, gratified Hitler. But in the long run, the effort to straddle failed; Germany's political and economic drive into the Balkans compelled Romania to accommodate. Their trade treaty of March 23, 1939—following immediately on Hitler's absorption of Romania's erstwhile ally, Czechoslovakia, into his Reich as the Protectorate of Bohemia and Moravia (see section 3)—recognized the Reich's predominance not only in Romania's foreign trade, but also in its general economic development, for which binational companies were to be organized. Yet, though it put Romania's raw materials at Germany's disposal, this and subsequent economic agreements were not ipso facto disadvantageous to Romania, as its agricultural surpluses were thenceforth absorbed at high prices by Germany and its industrial economy was modernized and rationalized by German capital and expertise over the next five years.

Some strain between Germany and Romania had been provoked when Carol arranged that the troublesome founder-leader of the Iron Guard, Corneliu Zelea Codreanu, be murdered by the police during the night of November 29/30, 1938. In view of the ideological affinity between the Nazi and the Iron Guard movements, this event predictably enraged Hitler and prompted several German dignitaries to return their Romanian decorations. But considerations of *raison d'état* prevailed on both sides. Indeed, to compensate Hitler for this provocation, Carol prudently became more accommodating in the trade negotiations. Combining flattery with self-service, he also aped totalitarian styles and trappings more assiduously. On December 15, 1938, the Front of National Renaissance was founded as the regime's monoparty, and on June 21, 1940, in a step that Carol advertised as an advance from the corporate to the totalitarian state, it was reorganized as the Party of the Nation, with membership compulsory for all holders of public and corporate office and with the Iron Guard's now rehabilitated remnant quite prominent. But all this rigmarole never developed any authentic dynamism.

Carol's relations with the post-Codreanu Iron Guard make for intriguing reading. Initially demoralized, the Guard's élan was revived by Hitler's blitzkrieg in Poland, and on September 21, 1939, it avenged Codreanu by assassinating the royal prime minister. A retaliatory massacre of Guardists and others immediately ensued; but in early December, Carol dismissed his police and secret-police chiefs, allegedly for having failed to protect the assassinated premier but in fact as a gesture of appeasement toward the Iron Guard and Hitler. Several hundred imprisoned Guardists were then quickly released in January 1940, and another thousand or so permitted to return from German asylum in the course of the spring, on condition that they enter Carol's monoparty. Among these returnees was Horia Sima, who had succeeded to the movement's leadership through the violent attrition of all other candidates. He was now to be alternately arrested and courted by Carol and was to join the royal cabinet and then to denounce the king, as the Iron Guard and the regime groped toward a nervous and brittle détente. On balance, however, Carol's was the stronger hand, and he could have controlled the Guard had not a series of fiascos in foreign policy in the summer of 1940 utterly discredited him among all sectors of the Romanian public.

In the aftermath of Hitler's destruction of France in May 1940, which was a psychologically devastating blow to a Romanian elite always self-consciously Latin in culture, Stalin decided to collect the pledges secretly given him the previous August by Hitler in return for Soviet neutrality at war's outbreak. On June 26, 1940, he issued an ultimatum to Romania to cede Bessarabia and the northern half of Bukovina to the Soviet Union. The Romanian Crown Council was advised by Berlin to comply and did so two days later. On July 1, 1940, Romania renounced an Anglo-French guarantee that dated from April 13, 1939, and three days later, Carol appointed a pro-German premier who promptly took Romania out of the League of Nations and announced its wish to join the Axis camp.

Such desperate endeavors to purchase German protection against further revisionist demands on Romanian territory came too late. At the spectacle of Romania's yielding to Moscow, the Hungarian and Bulgarian governments decided to present their claims as well. The latter's was settled relatively easily with the retrocession to Bulgaria of Southern Dobruja, whose population was admittedly almost 80 percent non-Romanian, under an agreement concluded on August 21 and formalized by the Treaty of Craiova on September 7, 1940. But the Hungarian demand for Transylvania failed of a bilateral accommodation and was then arbitrated by Germany and Italy on August 30, 1940, with Hungary regaining approximately two-fifths of its 1919 territorial losses to Romania (see section 4). Altogether, these four amputations cost Romania one-third of its territory, two-fifths of its arable area, more than two-fifths of its forests, most of the lands on which promising new industrial crops were being cultivated, and one-third, or over 6 million, of its people, of whom about half were ethnically Romanian.[12]

These losses were more than even the generally docile Romanian people were prepared to forgive Carol. For a decade, he had extracted painful political and economic sacrifices from the society in the name of national defense, and over the past several months, he had repeatedly sworn that "not a single furrow" of "eternal" Romanian territory would be yielded. And now he had supinely allowed his country to be truncated without a shot being fired in its defense. The neighboring Polish regime, which had also extracted sacrifices and emasculated parliamentary government in the name of national mobilization, had at least fought and thus sustained national honor in its own extremity the previous year. Execrated by the entire nation and pursued by the bullets of the Iron Guard—which, true to its spasmodic record, now bungled a belated attempt to apprehend him—Carol fled into exile on September 6, 1940, first to Spain, then to Latin America, and finally to Portugal. He left his nineteen-year-old son Michael to reign over an army–Iron Guard duumvirate led by General Ion Antonescu and Guardist leader Horia Sima.

In this post-Carol duumvirate, the army, demoralized by its inactivity during the recent surrenders of national territory, initially appeared to be weaker than its Iron Guard partner, basking now in the heady atmosphere of open alliance with Nazi Germany. The Guard was declared the sole legal party of the National Legionary State proclaimed on September 14, 1940, and was assigned responsibility for the nation's moral and material regeneration. Adventurers and opportunists flocked to it. Starting on the second anniversary of Codreanu's murder, the Guard went on a sustained rampage, slaughtering scores of political opponents, massacring Jews, and hinting ominously at the coming "popularization" of the officer corps. This protracted radicalism, however, proved its undoing. On the eve of his Russian campaign, Hitler required a disciplined and stable Romania, productively harnessed to the German war effort. Accordingly, he allowed army chief Antonescu to suppress the ideologically sympathetic but turbulent Iron Guard in three days of ferocious street fighting, January 21 to 23, 1941. At the same time, he granted Sima and a few fellow Guardists political asylum in Germany and kept them in reserve as insurance for Antonescu's reliable behavior for the rest of the war. Thus, ironically, was an authentic fascist movement emasculated with Nazi connivance at the moment when Germany's power in Europe and over Romania stood at its zenith. As it had at the time of Carol's royal dictatorship two years earlier, the Iron Guard collapsed when confronted with a strong hand. By carrying its negative radicalism to uncompromising, and hence irrational lengths, it had thrown away repeated opportunities to achieve power. As a symptom of the ills of Romanian society and as the nemesis for the falsehoods of Romanian politics, the Iron Guard was a serious phenomenon; but as a remedy for these very ailments, it was trivial.

General Antonescu dismantled the National Legionary State on February 15 and replaced Iron Guard functionaries with military personnel. He elicited public endorsement in a lopsided (99.91 percent) plebiscite held on March 2 to 5, and took Romania into the war against the Soviet Union as Germany's most valuable and valued ally on June 22, 1941. Though originally a professional disciple of French military models, Antonescu's reading of the European balance had led him to recommend a German association for Romania well before the war. As a "political" general, Antonescu had also for a time cultivated the Iron Guard and, after serving as war minister in 1937 and 1938, had incurred Carol's displeasure by appearing as a character witness for Codreanu when the Iron Guard's founder was on trial for treason in May 1938. Furthermore, Antonescu's puritanical disdain for the corrupt royal camarilla, expressed by his probe of dubious armaments contracts awarded to a royal crony, had prompted his dismissal as war minister and transfer to a provincial assignment. But such "exile" for rare personal rectitude only enhanced his popularity. Early in July 1940, Antonescu had even been arrested by Carol, now desperately seeking to protect himself against the expected national outrage over his recent territorial cessions to the Soviet Union and realizing that the general was the logical beneficiary and leader of such a nationalistic backlash. But these desperate royal exertions were to no avail. On the morrow of the partition of Transylvania, Antonescu vaulted straight from detention to power while Carol slid from power into exile.

Courageous and incorruptible, but also peevish and egotistical (he spoke of himself in the third person), Antonescu had the support of a united public for his attempted and initially successful recovery of Bessarabia and northern Bukovina from the Soviet Union—a feat for which he was designated marshal by King Michael on August 23, 1941. And Romania's contribution to the Axis war effort against the Soviet Union, which Antonescu sustained even after Romania's own irredentas there had been recovered, was far more substantial than that of any other German partner. As a result, Antonescu became Hitler's favorite ally, the first foreigner to be awarded the Knight's Cross of the Iron Cross, the only one to be solicited for military advice, and probably the only person—German or foreign—permitted to out-talk and contradict the *Führer*. Within the Axis, Antonescu defended Romania's sovereignty and national interest with much stubbornness. During the triumphant first half of the war, he joined eagerly in the genocide of the Jews. When the Romanian public and elite became skeptical of a continued war effort after Stalingrad, which was a Romanian as well as a German military debacle, Antonescu replied that the recent recovery of Bessarabia and northern Bukovina would prove ephemeral unless the Soviet Union were categorically defeated, and that Romania could hope to recover the lost fraction of Transylvania only by outperforming Hungary for Hitler's favor and thus inducing him eventually to reverse the partition of 1940. The first part of his analysis proved correct, the second false, as it was to be the Allied powers that eventually returned all of Transylvania to Romania as a reward for abandoning Nazi Germany before Hungary did (see section 4).

During the early war years, the urban and industrial prosperity of the late 1930s finally spilled over into the villages and the agricultural sector. This was not due to any structural reform of the rural economy, but simply thanks to massive German procurements of Romania's agricultural produce and to the recruitment of surplus rural manpower into Romania's own army. During the war's last year, however, this brittle prosperity vanished as the countryside became an object of both German and Soviet looting. By then, Romanian public opinion—elite as well as popular—was ready for a radical political and diplomatic shift in its war posture.

This about-face was achieved through a *coup d'état* by the hitherto figurehead King Michael against Antonescu on August 23, 1944, when Soviet armies, having reconquered all of northern Bukovina and half of Bessarabia, were poised to overrun the Romanian heartland. Antonescu, too, was by now prepared to sue for an armistice, but not without loyally informing Hitler beforehand. But this would have aborted the enterprise. From the end of 1943, his foreign minister had sought to avert the need for the royal coup by initiating secret peace feelers through various Romanian legations abroad, especially those in Ankara, Madrid, and Stockholm. But Romania's reluctance to concede the Soviet-claimed territories and Antonescu's personal scruples about deserting Hitler protracted these contacts until they were overtaken by King Michael's decisive action, taken on his own responsibility with only the *pro forma* endorsement of the leaders of the long illegal National-Peasant, Liberal, Social Democratic, and Communist parties. The next day, August 24, the Germans belatedly produced Horia Sima as head of a puppet government in Viennese exile, and the *Luftwaffe* began a futile three-day bombardment of Bucharest, the Romanian capital.

Antonescu was arrested (and eventually executed on June 1, 1946), and an armistice was concluded with the Allied powers on September 12, 1944, which reconfirmed the Soviet–Romanian border established in 1940 but, in compensation, indicated the restoration of the Hungarian–Romanian border of 1920 to 1940. On August 25, Romania had joined the Allies against Germany, and on September 7, it declared war on Hungary. Having fielded 27 divisions and suffered 500,000 casualties (of whom 300,000 died) in Hitler's war against Russia, having fueled and fed his armies with its oil and grain, and having then pulled off the most decisive volte-face of the war—one that turned the *Wehrmacht*'s southeastern hinge and opened the Balkan Peninsula and the Carpatho-Danubian basin to the Soviet army—Romania was now to contribute another 27 divisions and suffer a further 170,000 casualties (of whom 111,000 died) in the final Allied campaigns against Hitler. Its military prowess on both sides, from 1941 to 1944 and from 1944 to 1945, was an impressive contrast to the spineless surrenders of 1940.

It is clear that the illegal and underground Romanian Communist party, which managed to organize a few acts of wartime sabotage but never a serious guerrilla or resistance movement, had little to do with any of these important events and developments. Thus it was appropriate, albeit ironic, that at war's end, Stalin awarded the Soviet Union's highest decoration, the Order of Victory, to King Michael. It is also ironic that the absence of the bulk of the Romanian army from the country during the winter of 1944/45, as it was fighting at the side of the Soviet armies through northern Hungary and into Slovakia, facilitated the seizure of political power at home during these crucial months by this Communist party, whose previous activities had been so minimal.

7

In the first half of World War II, Bulgaria achieved the unique diplomatic feat of fulfilling its maximum irredentist aspirations under German patronage while limiting its participation in the Axis war effort to a lower level than any other satellite. Thus Bulgaria recovered Southern Dobruja from Romania in September 1940 and annexed Yugoslav and Greek Macedonia and Greek Thrace in April 1941. Yet though grateful to Nazi Germany for these acquisitions, its government protected Bulgarian Jews (though not those of annexed Macedonia and Thrace) from Hitler's genocide and rejected his pressure to sever diplomatic relations with the Soviet Union and join his war against it, pleading the Bulgarian people's traditional pan-Slavic affection for Russia. Though Bulgaria permitted its territory to be used as a staging and transit area for the German invasions of Yugoslavia and Greece in April 1941 (without, however, participating directly in those invasions) and though it formally declared war on the Western Allies on December 13, 1941, after the bombing of Pearl Harbor, the actual military engagement of Bulgaria's armed forces was for long limited to relatively easy occupation duty in Yugoslavia and Greece. But from the winter of 1943/44 onward, Bulgaria's cities were periodically bombed by the British and the Americans, and for a brief period in September 1944, it found itself again in the unique and this time bizarre position of being at war with both the Allied and the Axis powers. At war's close, Bulgaria was permitted to retain Southern Dobruja, acquired from Romania, but had to disgorge the territories annexed from Yugoslavia and Greece.

The credit for Bulgaria's astute maximization of gains and seeming minimization of commitments and liabilities in the war's first half goes to Tsar Boris III, who had been the constitutional monarch since the end of World War I but had staged a royal coup against the party system on January 22, 1935. Though avoiding any ostentatious display of personal power as well as the trappings or rhetoric of the fashionable totalitarian models of that day, Boris was the decisive and controlling force in Bulgarian politics and policy from the morrow of his royal coup until his sudden death at the early age of forty-nine on August 28, 1943. Cultivating the image of a frugal, unassuming, and simple man who would rather indulge his hobby of driving locomotives than immerse himself in affairs of state, which he supposedly took on only reluctantly as a royal duty, Boris was in fact a ruthless, albeit pessimistic, operator and cool intriguer. Adept at playing people off against one another, he applied this skill not only to his own ministers, generals, and politicians, but also to Hitler and Mussolini. Neither intoxicated nor intimidated by Hitler, who repeatedly and correctly described him as a political fox,[13] Boris affiliated Bulgaria with the Axis, but declined collaboration beyond the limits of what he viewed as his country's own strictly defined interests. The absence of a *Volksdeutsche* minority in Bulgaria deprived Nazi Germany of a lever that was available in its other wartime East Central European client states for extracting further concessions.

Boris's policies appeared initially to be sound in a number of different, yet mutually reinforcing and concentric ways.

In the economic arena, his governments promoted the expansion of intensively cultivated fiber, oil-bearing, industrial, garden, and other specialized and highly profitable crops over grains, and invested generously in agriculture-related processing industries. Though the country's economy thus remained overwhelmingly agricultural, it was a qualitatively higher stage of agriculture—involving more intensive use of labor, land, and capital, and entailing higher levels of peasant prosperity—than characterized the economies of the other Danubian states, which limped between a precarious grain-based agriculture and prestigious but unsound heavy-industrial investments. In terms of both direct, palpable results and the less tangible but even more important factor of accumulated experience and know-how, these economic policies of the late 1930s and the war years, though initiated with Nazi German encouragement and under German tutelage, have stood Bulgaria in good stead even into the postwar Communist era.

The annexations of the Yugoslav and Greek parts of Macedonia and of Greek Thrace (with access to the Aegean Sea) were so popular with the Bulgarian public that they largely neutralized the pro-Allied forces that were opposed to Boris's foreign policy. Indeed, even the Bulgarian Communists were caught up in this national enthusiasm and sought to contest the legitimacy of Tito's Yugoslav Partisans operating in Macedonia. But the high-handed condescension and contempt for local particularism with which the centralizing Bulgarian authorities administered Yugoslav Macedonia soon soured the welcome with which they had originally been greeted there. In Greek Macedonia and Thrace, their behavior was even harsher, expelling and killing several thousand Greeks and thus refueling a legacy of bitter ethnonational hostility that still lingers. In yet another sense, too, the very popularity of these territorial acquisitions turned into a diplomatic liability during the second half of the war because it then constrained Bulgarian feelers for acceptable armistice terms from the British and the Americans, who were, of course, committed to the claims of their wartime Yugoslav and Greek allies for full territorial restoration.[14]

When Boris's government had first declared war on the United States and Great Britain in December 1941 to please Hitler, it could plausibly be presented as a low-risk, merely symbolic gesture. After all, British power seemed to be at a nadir, while America was remote and licking its gaping wounds in the Pacific. To propitiate Hitler with this seeming formality as compensation for refusing to participate in his war against the Soviet Union seemed to be a shrewd bargain. But after the final loss of the *Wehrmacht*'s power of initiative on the eastern front in the great tank battle of Kursk in July 1943, after which the westward advance of the Soviet armies became relentless and almost uninterrupted, and after the Western Allies' establishment of their domination of the Mediterranean at about the same time, especially with the British and American bomber attacks on Romanian oil fields and on Bulgarian cities in the following winter, the seemingly clever policies of 1941 turned sour and ominous. The sudden death of Boris in the late summer of the pivotal year 1943 compounded the political and diplomatic uncertainty that was prompted by these impressive feats of Allied arms.

Boris was succeeded by a three-man regency that initially appointed a pronouncedly Germanophile cabinet, Hitler's energetic response to the fall of Mussolini having won him renewed fear and awe in the capitals of his satellites. But over time, a pair of subsequent Bulgarian governments became increasingly lukewarm toward the Axis cause and eager to find a way out of the war while retaining the territorial annexations of 1940 and 1941. Simultaneously, various pro-Allied Bulgarian opposition parties and personalities recovered confidence and visibility. Among them were Western-oriented democratic figures who on September 2, 1944, belatedly formed a short-lived government committed to taking Bulgaria out of the war and the pro-Soviet Fatherland Front coalition. Since the latter was to be the ultimate beneficiary of the tumultuous events of September 1944, in the course of which it seized power, a brief description of it is in order.

The Fatherland Front was formally established in June and July 1942 and composed of minority sectors of the Agrarian and Social Democratic parties, of a leftist group of intellectuals and military reserve officers named Zveno (Link), and of the underground Communist party. But until the eve of its seizure of power in September 1944, it remained a rather hollow body, with each of its four components suspecting the others of being willing to make a separate power-sharing deal with the regime. Furthermore, it was for long a conglomeration of leaders without a mass following, for the Agrarian and Social Democratic majorities remained aloof from it (as did the pro-Western bourgeois parties), most reserve and active army officers rejected it as unpatriotic, and the Communist cadres themselves were being smashed and repressed during 1942 and 1943 by the efficient and brutal royal Bulgarian police apparatus.

Indeed, the Bulgarian Communist party—once the proud ornament of the interwar Comintern for its revolutionary militancy, its organizational solidity, and its mass following—had fallen on parlous days. Initially, its celebration of the German-Soviet Pact of 1939 to 1941 had briefly earned it greater indulgence at the hands of the royal regime (which also endorsed the pact) than was extended to Boris's democratic opposition. And it reciprocated for this governmental leniency by curbing the traditional militancy of its May Day demonstrations in 1940. Similarly, the Communists sought to cater to Bulgarian nationalism in the matter of the territorial acquisitions of 1940 and 1941, only to be trumped in the resultant competition for Moscow's favor by the Yugoslav Partisans' greater anti-German activism after Hitler's invasion of the Soviet Union in mid-1941. Now the Bulgarian Communists were caught by Moscow's demand that they, too, engage in guerrilla-type resistance to Boris's Axis-affiliated regime—a demand that they were unable to meet, instead resorting to occasional assassinations of secondary pro-German political figures and regime functionaries during 1943.

Though veterans and sympathizers of the more active Yugoslav and Greek Communist resistance movements often deride their Bulgarian comrades-cum-rivals for failing to rise to comparable levels of revolutionary militancy, it must in fairness be acknowledged that it is psychologically, morally, politically, and logistically far more difficult to prosecute a guerrilla-type war against an intact state machine and a functioning social order of one's own nationality (as the Bulgarian Communists were called on to do) than to fight such a war against the occupiers of one's country in conditions where the apparatus and the authority of the state have disintegrated and much "loose" war material is available (as was the situation in wartime Yugoslavia and Greece). Despite these obstacles, the Bulgarian Communists did manage in 1944 to organize a modest degree of guerrilla activity—formally under Fatherland Front auspices—but it was utterly insufficient to defeat the Bulgarian army or topple the government without the assistance of an Allied army. And thanks to the strategic decisions of the Big Three, this meant, in practice, the Soviet army.

The craving to retain the Macedonian and Thracian annexations, which Bulgarians regarded as rightfully and justly theirs, as well as the fear of German retribution à la Italy in the summer of 1943, had lured Bulgarian governments to drag their heels in the matter of leaving the war until the last days of August 1944, when Romania's about-face the previous week forced Bulgaria's hand and ended a long period of suspended diplomatic animation. Bulgaria dispatched envoys to Cairo to negotiate armistice terms with the Western Allies, with which it was at war. Though legally correct, it was politically unwise thus to ignore the Soviet Union, whose armies were approaching Bulgaria's northern doorstep, whose accredited diplomats were resident in Sofia (the Bulgarian capital), and whose mediation it would have been prudent to solicit. Moscow now improvised quite deftly. Even though (or, perhaps, precisely because) the "neutral" Bulgarian government of June to August 1944 was replaced on September 2 by a categorically pro-Western one, which was committed to accelerating the armistice process with the British and Americans, the Soviet Union declared war on Bulgaria three days later (September 5), thereby juridically ensuring itself a decisive voice in that process. The Bulgarians responded by petitioning the Soviets for an armistice (September 6) and declaring war on Germany (September 8). The Soviet army, which had been pouring into Bulgaria since the morning of September 8 and meeting no resistance, surprisingly agreed later that day to the Bulgarian armistice petition, specifying that a cease-fire go into effect on September 9—even though this step entailed formally acknowledging and thus implicitly legitimating a "bourgeois" government that was due to be overthrown by a Fatherland Front *Putsch* also on September 9. Perhaps the Soviet military authorities knew nothing of the Fatherland Front's (and thus the Communist party's) political plans, perhaps they deemed the Front too weak to seize power, perhaps they did not care, perhaps their focus was preponderantly on military priorities to the neglect of political objectives, or perhaps the armistice was meant to throw sand into the eyes of the doomed government.

Whatever the explanation, the vengeance that the Fatherland Front now wreaked on its political rivals was to be particularly savage, making no distinction between pro-Westerners and pro-Germans. Thousands of old scores were settled, and the proportion of the population executed was higher than in any other former Axis state—despite Bulgaria's level of participation in Hitler's war having been the lowest and having entailed the least sacrifices of any Axis partner. As with neighboring Romania, the Germans now vainly sought to form a puppet Radical Right Bulgarian government-in-exile. In another parallel with Romania, the Bulgarian army was now attached to the Soviet campaign to clear the *Wehrmacht* out of southeastern Europe, in the course of which it fought through to Vienna, suffering 32,000 killed out of 340,000 troops engaged. Meanwhile at home, the Communist party was consolidating its grip on power, a process that (as in Romania) was facilitated by the army's—especially the officer corps'—absence from the country.

8

Albania, though formally an independent state, had become an economic and diplomatic client of Italy in the mid-1920s. This Italian hegemony was acknowledged by the other European Great Powers and was initially deemed useful by the royal Albanian government of King Zog as a deterrence to Yugoslav and Greek aspirations to partition Albania between them. Furthermore, Mussolini, who viewed Italy's stake in Albania as basically political and strategic rather than economic, was undoubtedly generous (though hardly altruistic) with investment and financial assistance to Albania. King Zog's problem was to strike a balance between, on the one hand, his and his country's need for Italian patronage and capital and, on the other, the restiveness of his own people, particularly the new intelligentsia of this late-modernizing society, which resented the Italian hegemony as a national humiliation.

By the early 1930s, Zog was resisting further Italian encroachments, but his search for alternative Great Power support and protection drew a blank, as neither Paris nor London nor Berlin was prepared to offend Rome over what was to them a tertiary country. They advised Zog, who was standing up to the severance of Italian subsidies and to the menacing Italian naval demonstrations in Albanian waters, to capitulate. But by now the Albanian elite's nationalistic sensitivities had to be reckoned with, and they reinforced his own reluctance to revert to the role of an Italian client. In 1935 and 1936, accordingly, Zog somewhat liberalized his authoritarian regime in order to compensate himself with enhanced domestic support for his current international isolation.

Mussolini's mid-decade preoccupation with Austrian issues, then with his Abyssinian campaign, and finally with the Spanish Civil War gave Albania and Zog an extended and heady, but ultimately illusory, reprieve. A temporary compromise was reached in 1936. Its terms, though confirming a substantial Italian stake in Albania's economy and administration, were something of a defensive victory for Zog, considering the lopsided imbalance of power between the two protagonists. But as soon as Mussolini was free from the other international claims on his attention, he moved swiftly and ruthlessly. On March 25, 1939, he delivered an ultimatum demanding a formal Italian protectorate over Albania and the stationing of Italian garrisons on its soil. When Zog stalled, his country was swiftly bombarded, invaded, and occupied between April 7 and 10, 1939, and he was driven into exile.[15]

The precipitating stimuli to Mussolini's sudden and peremptory crackdown appear to have been concern lest a new Yugoslav regime repudiate the preceding one's acknowledgment of Italian hegemony over Albania, and a wish, born of wounded vanity, to emulate Hitler and at the same time repay him in kind for having occupied the rump Czech state on March 15, 1939, without consulting his Axis partner. A similar motive also prompted Mussolini's invasion of Greece from Albania on October 28, 1940, under the pretext of recovering Albanian, not Italian, irredenta in Greek Epirus. The Greek army soon turned the tables, however, first halting the Italians and then pushing them back into southern Albania. Italian military morale never recovered from this fiasco, which in April 1941 precipitated the German invasion of the Balkans as an operation to rescue the embattled and weakening Italians. The Albanians had initially applauded the victories of the Greeks but then changed their attitude when Athens signaled its (eventually unrealized) intention to annex southern Albania (northern Epirus to Greek nationalists). At the conclusion of the April 1941 Balkan campaign, the Germans almost doubled the territory and population of Albania—nominally an Italian protectorate—by assigning to it parts of occupied Yugoslavia and of Greece with substantial ethnic Albanian populations.

This was to be reversed at war's close, though the issue continues to irritate Albania's relations with its neighbors.

The flight of Zog had initially left the Albanians leaderless and bewildered, as he had not permitted political parties to be organized or groomed any alternative leaders during the fourteen years of his modernizing but authoritarian rule. Thus the Italians at first had little difficulty in finding Albanian collaborators. Soon, however, two, then three, competitive anti-Italian resistance camps emerged. The first was spawned among the north-central clans and led by Abaz Kupi, the only officer of Zog's gendarmerie who had tried to repel the Italian invasion in April 1939 and who in 1940 began logistic and political preparations for a guerrilla rising on behalf of the exiled Zog. Two years later, Kupi merged his band into the broader National Liberation Movement, which had been launched in September 1942 as a broadly based, patriotic, popular resistance front by a small nucleus of Albanian Communists under the guidance of Tito's analogous Yugoslav Partisan movement. (The nuclear Communist party of Albania had itself been founded as recently as November 8, 1941—also under Yugoslav tutelage.) Finally, two months later, in November 1942, was organized a resistance movement of republican, anti-Communist nationalists who before 1939 had been in opposition to the "feudal" Zog. It called itself the National Front (Balli Kombëtar). Though pro-Allied and hostile to the Italian occupation, the National Front was committed to retaining the territorial annexations of 1941 from Yugoslavia and Greece.

The contrasting military and strategic approaches that were adopted by Tito's Partisans and Mihajlović's Četniks in Yugoslavia (see section 5) now found analogous expression in Albania. The Communist-led National Liberation Movement prosecuted its guerrilla war against the occupiers regardless of the costs to the population in terms of Italian retaliation, while the National Front husbanded its resources and avoided provocations that risked destructive reprisals. The latter also perceived its rival as a mere tool of the Yugoslav Partisans and hence as ready to betray ethnic Albanian lands to them. In one important political dimension, however, the analogy does not hold, for whereas the Yugoslav Četniks sought the restoration of the interwar royal regime and social system, the Albanian National Front anticipated profound postwar structural and social changes in a liberal republican direction.

The fall of Mussolini and the withdrawal of Italy from the war in the summer of 1943 crystallized these Albanian tensions. Both rival resistance camps now came into possession of substantial quantities of Italian arms. The Germans quickly introduced just enough *Wehrmacht* units to secure the few main towns and the Albanian coast against a possible Anglo-American landing, but otherwise made no political demands. Indeed, they reendorsed the Albanian territorial extensions of 1941. This astute German conciliation impressed Albanian nationalists and seduced the National Front into turning, in effect, from resisting the Italians to collaborating with the Germans. Thereupon the Communist-energized National Liberation Movement launched a ferocious civil war against it in the name of the overall Allied struggle against the Axis, including the repudiation of Axis-initiated wartime territorial transfers. This, however, provoked Abaz Kupi to secede from the National Liberation Movement in protest against what he perceived as its betrayal of Albanian national unity and national interests. In late November 1943, Kupi founded his pro-Zog Legality Movement among the north-central clans as an alternative to the two other camps. Nevertheless, the National Liberation Movement, backed by Tito's Partisans and able to claim anti-Axis purity, went on to win the civil war as well as the national resistance struggle, and thus to capture control of the country, recruiting and indoctrinating widely in the process—especially among youths, women, professionals, artisans, workers, and tenant-peasants.

Thus in Albania, as in Yugoslavia, the war of national liberation was a dynamic threefold process, entailing (1) internal political polarization, (2) a sense of national consciousness even among hitherto parochial and nonpolitical social segments, and (3) a radical upheaval of elites. And in exact imitation of Tito's pattern of escalating his AVNOJ from a national liberation movement into a provisional government at the Bihać and Jajce congresses of November 1942 and November 1943 (see section 5), his Albanian Communist protégés elevated their National Liberation Movement at a pair of congresses held in the provincial towns of Përmet and Berat on May 24 to 28 and October 20 to 23, 1944. Throughout that year, the tutelage of the Albanian by the Yugoslav Communists increased as the former extended their control over the country and finally conquered governmental power and the national capital, Tirana, in the wake of the *Wehrmacht*'s withdrawal northward in November 1944.

Now, however, an unexpected denouement that was definitely not in the Titoist script was played out. After the elimination of all its domestic rivals, the National Liberation Movement (renamed the Democratic Front in 1945 after becoming the country's government), together with its controlling Communist nucleus, was wracked by a severe inner crisis over how to behave toward its Titoist mentors, to whose ideological and organizational abilities it owed its founding, its evolution, and its conquest of power. One faction, led by Koçi Xoxe, who was both organization secretary of the Communist party of Albania and minister of the interior (and thereby controller of the police apparatus), advocated the acceptance of "comradely" Yugoslav hegemony over Albania to the point of entering the Yugoslav federation. For a while, his position and policy appeared to be ascendant. But the eruption of the rift between Tito and Stalin in 1948 enabled his rival Enver Hoxha, who was the Communist party's secretary general and the government's prime minister, to turn the tables and exploit Soviet support both to unhinge the Yugoslav grip and to eliminate his domestic rivals within the party. An additional bonus accuring to Hoxha from this repudiation of the overbearing Yugoslavs was that it enhanced his popularity with the Albanian people, who viewed it as partial compensation for his regime's embarrassing earlier return, in 1945, of the Yugoslav territories that the Germans had assigned to Albania in 1941 in recognition of their substantial Albanian populations. In the subsequent vituperation over the antecedents and causes of this rupture of 1948, the Albanians charged that Yugoslavia had inflicted imperialistic vassalage on a poor and weak Albania between 1944 and 1948, while the Yugoslavs protested that they had altruistically incurred many sacrifices to help the Albanian ingrates.

9

In recent years, scores of books and articles have addressed the question of how and why the Western Allies, despite their tremendous contributions to winning World War II, apparently "lost the peace" to the Soviet Union. This is not a suitable point for a review of that literature or for any general contribution to that discussion. Nevertheless, two limited and specific comments are in order. First, as regards the postwar fate of East Central Europe, momentous consequences ensued from the failure of the Western Allies to take effective advantage of the fall of Mussolini and the capitulation of Italy in the summer of 1943 (respectively, on July 25 and September 8). Had they done so, the governments of several smaller Axis partners in East Central Europe would probably also have been emboldened to risk breaking with Nazi Germany at that time, when the Soviet armies were still engaged far to the east, deep inside the Soviet Union. (They retook Kharkov on August 23 and Smolensk on September 28, 1943.) Instead, Hitler's prompt occupation of Italy, contrasting so vividly with the Western Allies' sluggishness, effectively served to discourage the others. The governments of his East Central European satellites and dependents thereupon resubmitted themselves to his domination for another year, from mixed motives of fatalism and conviction.[16] By the time they again steeled themselves to hazard leaps out of the Nazi German embrace in the summer and autumn of 1944, the Soviet presence was on them. And the Soviet rulers did not deem it in their own interest to permit the recovery of effective East Central European political independence, such as a Western Allied presence might have authorized and protected under a militarily different war-termination scenario.

The second, related, point is that the hegemony over East Central Europe that was achieved through the Soviet Union's immense sufferings and sacrifices, its military prowess, and its diplomatic exertions during World War II has since become a powerful justification of the Soviet system in the eyes of its own elite and public. That hegemony has been the most visible and palpable prize of the great Soviet victory and therefore continues to function as a powerful moral bond between the regime and its peoples and among the various sectors of the Soviet elite. Hence it probably cannot be relinquished without jeopardizing that internal Soviet legitimacy for which it is both catalyst and keystone. And these considerations will override Soviet awareness that the East Central European objects of the hegemony are sullen and that its objective blessings even for the Soviet Union itself are mixed.

3

The Communists Come to Power

1

To stipulate that their hegemony over East Central Europe is deemed vital and nonnegotiable by Soviet leaders does not, in and of itself, indicate the precise forms and structures through which that hegemony is to be exercised. Even to go one step further and note that their close brushes with military catastrophe in 1941 and 1942 have left the Soviet leaders (beginning with Stalin in his own time) obsessed with security concerns and with a propensity toward military definitions of their system's security also does not ipso facto explain the style of the Soviet Union's imposition of its control over East Central Europe after World War II. After all, geomilitary security could have been readily ensured by means other than the imposition of Communist regimes and the attempted *Gleichschaltung* (enforced coordination) of socioeconomic arrangements throughout postwar East Central Europe to the model of the Soviet Union itself. To account for the methods that Stalin and his heirs selected to operationalize Soviet hegemony over the area requires the introduction of ideological, systemic, contingent, and even idiosyncratic explanatory variables in addition to postulating "objective" security concerns.

Today we recognize that many of the Western academic analyses of the 1950s and 1960s subscribed to exaggerated images of a rigid blueprint that supposedly guided Moscow and the local East Central European Communists in implementing the procedures and arrangements that Stalin eventually selected to give effect to his perception of Soviet hegemonial requirements. But while validly correcting those earlier errors, we should not throw out the baby with the bath water. For though there probably never was a rigid, uniform Stalinist plan to be imposed in cookie-cutter fashion on every state and society of East Central Europe without regard to diverse national histories, institutions, and complexities, the general overall similarities in Soviet and local Communist behavior throughout the area in the first decade after World War II strongly suggest a unified conceptual political framework—more flexible at the beginning than toward the close of that decade, never entirely absent yet also never absolutely rigid.

W. Averell Harriman, the nonacademic and reflective American ambassador to the Soviet Union, suggested in April 1945 that Stalin was then seeking a way to finesse three alternative policy options: (1) extending the wartime alliance into continued cooperation with the United States and the United Kingdom; (2) establishing a tight Soviet security zone in East Central Europe, an area that had, after all, served as the springboard for Hitler's recent (and for many earlier) invasions of Russia; and (3) penetrating Western European (and other) societies and subverting their governments through the instrumentality of their Communist parties.[1] While it is clear in retrospect that these three options were incongruous and that the manner in which Stalin was soon to implement the second of them would abort the other two, it may well be that at the time, at war's close, such incongruity did not appear to be tantamount to utter incompatibility. The concept of "people's democracy," which Stalin authorized for East Central Europe as an intermediate social and political order ostensibly alternative to Sovietization and "proletarian dictatorship," may initially have been intended seriously, albeit experimentally, as a formula and an arrangement to harness the three options together, to implement the Soviet Union's perceived security needs and political interests in East Central Europe without sacrificing either the Big Three alliance or the potential attractiveness of Communism elsewhere. Unless we allow for such a latitudinarian conceptual possibility, certain anomalies in several East Central European states between 1945 and 1950 (as well as manifest Soviet unpreparedness and improvisation toward many of their bilateral quarrels) are difficult to explain on the alternative premise of a predetermined intention uniformly to Communize the area. Yet given the kind of person that Stalin was, the kind of operational system that Leninism-Stalinism had become, and the vast disparity in power between the dominant Soviet Union and the subordinate states of East Central Europe, the Soviet self-restraints that would have been required to realize such a scenario of harnessing the three options together by rendering people's democracy into an authentic, viable alternative short of monolithic Communist power were, alas, to be in short and precarious supply.

The logic of Stalin's conceptual stance and political orientation toward East Central Europe presumably entailed the following sequential considerations: (1) Merely denying the area to Germany or any other potential enemy of the Soviet Union would not suffice, for it was too weak to resist future pressure from such an inimical power; (2) Hence its regimes must be positively supportive of the Soviet Union, rather than merely uncommitted (let alone unfriendly); (3) Such a positive relationship with the Soviet Union could be reliably ensured only through some structural transformations in the "bourgeois" and "feudal" societies of the area, and not through the merely "superstructural" policy promises of even the most benignly inclined local bourgeois politicians (for example, Edvard Beneš); (4) The institutional form of these structural transformations would be people's democracy, a social form transitional between bourgeois democracy (the West) and mature socialism (the Soviet Union); (5) Since the people's democracies of East Central Europe were thus by definition placed in a less advanced historico-developmental niche than the Soviet Union, their subordination to it could be justified on ideological grounds and not merely by reference to raw power considerations; (6) By the same token, since they were also stipulated to be more advanced than the bourgeois systems, any potential backsliding on their part away from the Soviet Union and toward the West could be conveniently vetoed on ideological grounds as ostensibly retrogressive; (7) Though defined a priori as morphologically less mature than the Soviet Union, the people's democracies were to function as suppliers of capital to facilitate the recovery of the war-devastated Soviet economy.[2]

To the extent that this rationale for the people's democracies was intended (among other purposes) to straddle and finesse the triadic dilemma that Harriman had identified, it failed. Its implementation in East Central Europe required methods that were so harsh and so transparently dictated by Moscow as to arouse the alarm of the West, to damage the appeals of Communism elsewhere, to provoke deep resentment among the subject peoples, and eventually even to evoke "national Communist" umbrage within the Communist cadres of the people's democracies themselves against the substance and style of Soviet exploitation and domination. We now turn to comparing and contrasting this process in the several states of the area.

2

Social scientists tend—by and large, validly—to impute the causation of specific political developments to general social and secular trends. Thus they correctly note that wars often precipitate and accelerate profound political changes in the belligerent states. In reference to Poland, however, a note of caution is in order. While the military catastrophe of September 1939 and the subsequent years of destructive occupation and national resistance did open up Poland for deep political transformations away from its interwar system by lacerating its socioeconomic structure and radicalizing public opinion, these wartime changes did not point spontaneously, still less inevitably, toward a Communist solution. To impose themselves on Polish society by seizing and consolidating political power required a tenacious struggle by the Polish Communists and their Soviet patrons.

An inventory of Communist assets and liabilities in this struggle would have to include some variables of ambivalent value. For example, the Communists' identification with Russia—a historical national foe and now officially atheist to boot—was manifestly a drawback in their efforts to win acceptance among the Poles. Yet this same Soviet Russia was the only Great Power that categorically guaranteed Poland's valuable postwar territorial acquisitions from Germany (see Chapter 2, section 2). Somewhat less ambivalent and relatively more a Communist asset was the wartime decimation of the prewar gentry class and of the traditional intelligentsia, which drained the whole society's capacity for further civic resistance. Yet the net political benefit of these new social gaps and political openings might more plausibly have accrued to the peasant movement had Soviet power not backed the local Communists. Even the sharp rise in membership in the Polish Workers party (the Communists' formal new name) immediately after the war—from 30,000 to 210,000 between January and December 1945 and then to over 500,000 by January 1947[3]—was rather ambivalent in its political significance; many of the recruits were opportunists and careerists, while many others joined out of a sense of foreclosed fate, a feeling that there was no other alternative to rebuilding Poland. Both types lacked true ideological conviction.

A number of unambiguous liabilities weighed on the Polish Communists in the immediate postwar years. Unlike their Czechoslovak, Hungarian, Yugoslav, or Bulgarian counterparts, they lacked any leaders of genuine popularity or vivid achievement. The leaders that they did have were strangers to one another, some having spent the war years under German occupation, others in Soviet exile. Their major political competitors were free from the taint of collaboration with the Nazi German occupiers and, indeed, had led the major national resistance movement during the war, compared with which the Communists' own had been rather puny.

But the Communists were buoyed by some clear assets, which over the long haul outbalanced the liabilities and the ambivalent variables. Their Soviet sponsors were militarily and administratively present and active, while their competitors' British and American patrons were distant and inert; indeed, the real *virtu* of the Soviets and their local Communist protégés lay not so much in the sheer presence of Soviet military might as in their appreciation of the political leverage that it conveyed. Through their control of the Ministry of Security, the Communists effectively monopolized the state's instruments of internal force and violence. Through their control of the Ministry for the Regained Territories, they monopolized an extensive patronage apparatus for the distribution of the newly annexed lands, from which most of the German population fled or was expelled, to their nascent clientele. Their control of the extremely rapid and supposedly "spontaneous" process of distributing the lands and assets of large agricultural estates throughout Poland among the peasantry served a similar purpose and helped them to undermine the rival Peasant party, which, as a result of the destruction or neutralization of other potential loci of legal opposition, emerged as the keystone to any possible political resistance to the Communists.

Before turning to this pivotal confrontation between the Communist and Peasant parties, it is necessary to understand the demographic, economic, and domestic political significance of the paired loss of Poland's eastern borderlands to the Soviet Union and gain of western and northern territories from Germany. Allusions to the diplomatic consequences of this westward movement of Poland in terms of its subsequent dependence on Soviet support against eventual German revanchism have already been made.

This pair of territorial shifts and their associated population transfers (together with the wartime extermination of Polish Jews) transformed Poland from a multinational state, in which one-third of the interwar population had consisted of ethnic and religious minorities, into a nearly homogeneous Polish and Roman Catholic nation-state. They also opened its almost landlocked interwar geophysical profile via a new Baltic coastline of approximately 500 kilometers (310 miles) with several good harbors. And they rendered the spatial configuration of the state far more compact by sharply shortening its borders. Though the area gained from Germany (102,985 square kilometers or 39,752 square miles) is substantially smaller than the area lost to the Soviet Union (180,000 square kilometers or 69,480 square miles), it is economically far more valuable, containing the Silesian industrial and mining complex and, on balance, better farm land. Thus the overall bottom line of these territorial and demographic transfers is so manifestly favorable that considerations of sheer patriotism and *raison d'état* have linked the Polish nation to the regime that presided over them, resented though that regime is on many other grounds and regretted though its consequent security dependence on the Soviet Union may be.

It will be recalled that the postwar government consisted of a Soviet-sponsored trunk onto which a few "London" Poles had been grafted at British and American insistence at the close of the war (Chapter 2, section 2). The trunk was composed of a Communist core to which were attached Socialist and Peasant party splinters as well as some insignificant minor parties and a nominally nonpartisan but pro-Communist defense minister. The Socialist and Peasant splinters who affiliated themselves with the Communists did not do so with any of the apocalyptic enthusiasm that characterized their Czechoslovak analogues (see section 3), but out of a sense of ineluctable fate. Hence their attitude toward their Communist partners was occasionally skittish, though never openly disloyal. The returning "London" Poles were led by a former prime minister of the wartime government-in-exile, the authentic Peasant party leader Stanisław Mikołajczyk.

Apparently hoping that Poland's foreign-policy alignment with the Soviet Union would not necessarily require its internal political or socio-economic transformation based on the Soviet model, Mikołajczyk decided on a vigorous test of strength with the Communists. His hope was rendered superficially plausible by the contemporary example of Finland; but he overlooked the fact that Finland was then only peripheral to Soviet perspectives and expectations, whereas Poland, alas, was central. Mikołajczyk's political mettle was also stoked by his awareness that its genuine popularity with the peasant masses, as well as its status as the only secular political force authentically independent of the Communists, rendered his Polish Peasant party the natural magnet for all Poles ready to assert their opposition to Communist control of their country. And, indeed, these hopes and expectations were given a brief fillip; by January 1946, six months after Mikołajczyk's return to Poland, his party's membership had swelled to 600,000—outstripping that of the Polish Workers party, which, indeed, had suffered an absolute dip in membership during an interval in mid-1945.[4] The Communists and their associates were sufficiently chastened to postpone repeatedly the early and free elections to which they had supposedly committed themselves before and during the Potsdam Conference as a condition for obtaining British and American endorsement as Poland's government.

Such statistics and such popularity had, however, little true political significance. Though the Communists formally headed only six of the twenty-one ministries, they controlled most of the others through their deputy ministers or their splinter-allies. And through their direction of the security and police agencies as well as supposedly spontaneous worker "actions" and riots, they generated an atmosphere of intense intimidation, forcing the Peasant party, its activists, and its members to hover precariously between legality and illegality, ever vulnerable to the criminal code and even to sheer terrorization. Thus Mikołajczyk's intended test of strength was soon beaten into a rear-guard action, which he conducted with more courage than skill.

This ominous tension within the government between its Communist-dominated trunk and its Peasant branch was snapped as the result of a Socialist overture. On the one hand, acquiescent to the reluctance of their Communist allies to risk early and free elections and, on the other hand, concerned that the government obtain some sort of public ratification (so far, it was only the creature of Big Three fiats), the Socialists proposed a referendum in lieu of elections in which the voters would be asked to endorse the abolition of the interwar Senate, the current distribution of agricultural land to the peasants and the nationalization of heavy industry, and the new Baltic and Oder-Neisse frontiers—that is, the northern and western territorial acquisitions. The Communists endorsed this clever Socialist proposal, and the referendum was scheduled for June 30, 1946.

With the electoral route blocked by his nominal government partners-cum-political enemies, Mikołajczyk seized on the referendum as a device to demonstrate his popularity, his independence, and his leverage. Though the Peasant party had traditionally opposed parliamentary upper chambers—including the interwar Senate—as elitist dilutions of democracy, he now reversed this principled stance and asked his supporters to vote against the first of the three propositions, while joining the other ministers in calling for ratification of the other two. This questionable decision prompted a small but significant secession from his party. Also of dubious tactical wisdom was an Anglo-American decision to try to strengthen Mikołajczyk's hand by suspending credits to Poland during the referendum campaign, ostensibly as a protest against the repeated postponements of the pledged free elections.

It took ten days for the government to publish the official referendum results. It claimed positive endorsement of the three propositions by the following percentages: abolition of the Senate, 68; land distribution and industrial nationalization, 77.2; new frontiers, 91.6. But there exists persuasive evidence that these alleged results are spurious and that, despite blatant chicanery, provocation, and intimidation, a large majority of voters had actually rejected the first proposition—not from any sentimental attachment to the Senate but as a gesture of protest. The other two propositions were, of course, uncontested. Nationalization of industry was a relatively uncontroversial issue in the Polish tradition; in the 1930s, state capitalism had been more extensively developed in Poland than in any other European country except the Soviet Union and, possibly, Sweden. And the third proposition appealed automatically to reflexive patriotism.

Their management of the referendum campaign and its formal results left the Communists and their allies confident that they could now either win or successfully rig parliamentary elections, which were accordingly and at long last scheduled for January 19, 1947. The Peasant party ran against a coalition terming itself the Democratic Bloc and composed of the Communists, the Socialists, and two smaller parties. The Peasant group that had broken with Mikołajczyk six months earlier over the Senate issue in the referendum put up a nominally independent slate, but in effect supported the Democratic Bloc, as did yet another minor party. The campaign was characterized by escalating terror. In ten out of the country's fifty-two electoral districts, which contained about one-quarter of the population and where the Peasant party was traditionally strong, its candidate lists were disqualified. Almost 1 million other voters were disfranchised on the allegation of wartime collaboration with the Nazi German occupiers. Many Peasant party candidates, functionaries, and poll watchers were arrested and/or beaten. "Voluntary" open voting and fraudulent ballot counting were widespread. British and American protests were ignored.

The official electoral results were predictable. The Democratic Bloc was declared to have won 80.1 percent of the votes; its two nominally independent minor supporters, 3.5 and 4.7 percent; Mikołajczyk's Peasant party, 10.3 percent; with 1.4 percent of the vote scattered. The

Peasant party's claim, based on an early sampling of untampered ballot boxes in 1,300 out of 5,200 constituencies, to have received 60 to 68 percent of the votes was to no avail. It was now relegated to a mere 27 out of 444 seats in the constituent parliament, with the Democratic Bloc allocating 394 to itself, 7 and 12 to its pair of nominally independent supporters, and scattering the remaining 4. Mikołajczyk and his party were excluded from the government and ominously tarred with the brush of maintaining contact with the illegal resistance groups that still held out in the forests (see Chapter 2, section 2). Fearing arrest and worse, he fled to the West with the assistance of the American ambassador to Poland on October 21, 1947, to be followed shortly by several colleagues.

In retrospect, it is difficult to decide whether Stalin and the Polish Communists never intended to allow free elections in Poland—their promises before and during the Potsdam Conference having been but dust in the eyes of the British and Americans—or whether the disappointing Communist performance in the relatively free Hungarian parliamentary elections of November 1945 (see section 4) prompted a decision to renege on what may have been initially sincere assurances about Poland. I lean toward the first hypothesis.

With the elimination of the Peasant party as a meaningful political force, the Socialists became a magnet for those Poles wishing to register legal opposition to the Communists. After all, during the interwar decades, the Socialists had been a major party of impeccable Polish patriotic credentials, while the Communists had been a small, illegal coterie generally suspected of being a stalking-horse for Soviet Russia's predatory intentions toward Poland. And immediately after Poland's liberation, many Socialists returning from the West, from concentration camps, and from German prison and labor camps had sought to reassert their party's distinctive identity in relation to, albeit not its former hostility toward, its current Communist allies. Indeed, in the summer of 1946, the Socialist leader Edward Osóbka-Morawski even publicly denied the Communists' claim to being the governing coalition's leading party and demanded more cabinet representation for his own. And until early 1947, the Socialists had far more members than the Communists (800,000, compared with over 500,000) and stronger trade-union support. Thus what had begun as a dependent splinter had achieved some real popular substance. But all this availed little against the Communists' raw power to set the rules of the postwar political game. They obliged the Socialists first to purge themselves of 150,000 excessively independent members during the first half of 1947, then to yield up 200 recalcitrant middle-level leaders for arrest on the charge of affiliation with the illegal resistance in May and June 1947, then to join in declining America's invitation to Poland to participate in the Marshall Plan aid program to rehabilitate Europe in July 1947, then to withdraw from the Socialist International in March 1948 (after the previous month's Communist coup in Czechoslovakia; see section 3), then to purge themselves again and repudiate their prewar leaders in September 1948, and finally to dissolve their party and merge it with the Communists to form the Polish United Workers party (PZPR) in December 1948. By this date, the Socialists' membership had been truncated to half the Communists' (450,000, compared with 900,000).

In insisting on this coerced organizational merger, the Communists impatiently rejected some interesting final Socialist pleas that even in a people's democracy two working-class parties are desirable to have "reciprocal control" and "to guarantee freedom." Of the eleven members of the merged PZPR's new Politburo, eight were Communists and three former Socialists, while of the restructured cabinet's seventy-four ministers and vice ministers, fifty-three were Communists.

One might reasonably ask why the Socialists had been so blind to the handwriting on the wall as to cooperate unflinchingly in the Communists' destruction of the Peasant party during the first two postwar years, thus leaving themselves morally and politically naked when the Communists eventually turned on them. The answers are multiple, though not necessarily convincing. Many Socialist leaders appear to have persuaded themselves that the choice in the June 1946 referendum and the January 1947 elections was between "reaction" and "working-class solidarity." This belief was rendered ostensibly plausible by the fact that Mikołajczyk's transformation of his party into a catch-all basin had indeed opened it to some reactionary elements. Also, historically no love had ever been lost between the Socialist and the Peasant movements. And the Socialists may have sincerely feared that protecting or even stabilizing Mikołajczyk might provoke direct Soviet intervention. Furthermore, the most experienced and astute of the interwar Socialist leaders either had not survived the war or had remained in the West. Finally, and conclusively, the postwar Socialist party apparatus was infiltrated and cowed by the Communists.

Ironically, just as the coerced merger of the Socialist with the Communist party approached its culmination, the latter was passing through a severe inner crisis, one that was heavily predicated on the alienation between those of its leaders who had survived the war years under German occupation in Poland and those who had spent them in the Soviet Union. Superimposed on this fault line were the profound stresses inflicted on all the Communist parties of East Central Europe by the rift between Tito and Stalin in 1948. At the risk of some oversimplification (but not of falsification), one might hazard the following generalization: the wartime "local undergrounders" were concerned to pursue a distinctive national (in this case, Polish) road to so-called socialism, whereas the "Muscovites" insisted on the closest possible imitation of their Soviet model and the most slavish deference to Soviet signals. This deep crisis, which wracked the entire Communist movement throughout East Central Europe, will be closely analyzed in Chapter 4; suffice it to note here that in Poland, as in most of the area's other countries, the "Muscovites" initially won but, unlike their analogues in the area's other parties, did not execute their defeated "local" rivals or humiliate them at show trials. Perhaps the fact that all Poles, including the Communists and especially those Communists who had sought Soviet asylum during the interwar and war years, had suffered so much at Stalin's hands accounts for the relative leniency of his Polish "Muscovite" satraps in the late 1940s. They were not willing to risk reopening the trap door to renewed intraparty bloodshed, lest some day they be pushed through it.

Until after the Peasant and Socialist parties were eliminated as autonomous political forces, the Communists behaved quite gingerly toward the Roman Catholic church. Indeed, the government and the church had worked in parallel to give a Polish character to the newly acquired ex-German lands, and the Communist chief of state, Bolesław Bierut, used to attend ecclesiastical ceremonies on special occasions in his official capacity. Though the regime—but not only the regime—regarded Pope Pius XII as having favored the Germans during the war and therefore denounced the Concordat of 1925 on September 16, 1945, it postponed a struggle with the Polish episcopacy until after such struggles had erupted in Czechoslovakia and Hungary. Even after the pope threw down the gauntlet by excommunicating Communists in general (not specifically mentioning Poland) on July 13, 1949, the Polish regime responded cautiously, albeit palpably. In September 1949, it nationalized the church's network of hospitals; in January 1950, it took the big church welfare organization Caritas under state control; and in March of that year, it confiscated church-owned landed estates, which had been untouched by the postwar land reform. The new Polish primate, Archbishop Stefan Wyszyński, thereupon bid for a truce, which was quickly arranged. On April 14, 1950, he and the government signed an agreement in which the church publicly repudiated the surviving underground resistance (when the Vatican was still recognizing the vestigial government-in-exile in London) and endorsed the regime's lively peace propaganda, while the state reciprocally authorized that religion continue to be taught in its schools and chaplains continue to function in the armed forces, jails, hospitals, and so on.

Pope Pius XII's refusal to regularize canonically the Polish church's administration in the former German lands, and his insistence on appointing only provisional Polish apostolic administrators rather than ordinary bishops there pending a general European peace conference to formally ratify the new borders, embarrassed Wyszyński and gave the Communist government an alibi to renew its pressure on the church.[5] It charged that the refusal of most priests to sign the Soviet-sponsored Stockholm Peace Appeal of June 1950 was a reneging by the church on the agreement signed in April. During 1953, it decreed that all appointments to ecclesiastical office required its approval, subjected a bishop and several priests to shaming show trials on charges of espionage and economic sabotage, quietly imprisoned many other clergy, and interned Wyszyński (who had been designated a cardinal on January 12) in a monastery, where he languished until the dramatic events of 1956. But it never put him on trial (in contrast to the Hungarian and Yugoslav governments' respective handling of József Cardinal Mindszenty and Archbishop Alojzije Stepinac) and utterly failed to dent, let alone break, Wyszyński's and the church's moral authority with the overwhelmingly Catholic Polish public. The regime's campaign to undermine him and his episcopal hierarchy through a movement of so-called patriotic priests was a derisive failure. On balance, the Polish Catholic church more than recouped its postwar material losses through its flock's renewed fervor. It even gained followers in addition to the traditionally faithful peasantry as other social groups that had been indifferent or even anticlerical gave it their allegiance as a mark of political and spiritual protest against Stalinist trends. Even Communist party members would regularly, if surreptitiously, participate in the church's liturgy and sacraments. It thus became the only national institution that managed to checkmate its attempted subordination by the Communist regime and to retain a strong autonomous role in public life.

The armed forces were yet another structured institution to be subjected to severe pressure and purges in the course of the Communist takeover. But in this case the initiative appears to have been directly Soviet and was so heavy-handed as to embarrass Poland's own Communists. It will be recalled that the bulk of the old Polish officer cadre that survived the September 1939 catastrophe remained loyal to the government-in-exile in London and continued to fight the Germans on various western fronts and in the underground Home Army. After 1943, the Soviet Union's Polish Communist protégés were thus obliged to sculpt a Polish army from the Kościuszko Division, various Communist partisan bands, and some coerced or voluntary recruits from the prewar cadres (see Chapter 2, section 2). At the end of the war, many of the Soviet officers who had been seconded to this Soviet-sponsored Polish army between 1943 and 1945 were recalled into the Soviet army and replaced by Communist-screened, and hence presumably reliable, Polish officers.

But with the formation of NATO on April 4, 1949, and of the Federal Republic of Germany (West Germany) on September 7, 1949, Moscow apparently decided that even this Communist-controlled and Communist-commissared Polish army was insufficiently reliable. On November 6 of that year, Marshal Konstanin K. Rokossovsky—one of the most talented wartime commanders of the Soviet army and a man of partly Polish genealogy—was suddenly imposed on Poland as defense minister and commander in chief of the armed forces, followed soon by many other Soviet officers. On May 10, 1950, Rokossovsky even became a member of the Politburo of the PZPR, which suggests the extent to which his responsibilities included political supervision as well as military professionalization in Poland. He replaced the armed forces' few prewar holdovers with Soviet officers and subjected several of the former to demeaning show trials, expanded the army and modernized its equipment, imposed universal conscription, and developed new mechanisms (for example, the Feliks Dzierżyński Academy) to ensure the reliable political indoctrination of the armed forces. Quite apart from these activities and policies, Rokossovsky was already anathema to the Poles as the wartime commander of the Soviet Army Group that had stood by passively as the *Wehrmacht* suppressed the Warsaw insurrection in the summer of 1944. Though he had obviously been following orders during that tragedy and though he now delicately retained traditional Polish uniform cuts, anthems, and other such symbols, Rokossovsky's very presence, let alone his current role, in Poland in the 1950s was universally resented. It reflected obtuse Russian insensitivity to Polish—even Polish Communist—national sensibilities and, superimposed as it was on the "Muscovites' " purge of the "local undergrounders" within the Communist party, indicated Stalin's definitive rejection of distinctive national roads to socialism.

3

Both the differences and the similarities between Poland and Czechoslovakia in the matter of the Communist assumption of power are significant. In Czech and Slovak, unlike Polish, collective memories, Russia was perceived as a historical friend, and the Soviet Union was not regarded as an accessory to Nazi Germany's destruction of the interwar state and occupation of the country. At war's end, the Czechoslovak government-in-exile, unlike the Polish one, was able to return to its capital from London, albeit via Moscow and with heavy Communist representation. Furthermore, just as the interwar Czechoslovak Communist party had been a major, legal, respectable, and indigenous one that consistently drew over 10 percent of the popular vote in free parliamentary elections, whereas its Polish counterpart had been small, intermittently outlawed, and widely viewed as a subversive agent for an external predator, so the postwar public response to Communist preeminence in the government was quite different in Czechoslovakia, where this new pattern was widely deemed to be legitimate even if regrettable, than it was in Poland. Also, while postwar Poland was constitutionally a centralized state, the political energies released by wartime Slovak independence and especially by the impressive Slovak uprising in the autumn of 1944 (see Chapter 2, section 3) extracted a semifederalist restructuring of the Czechoslovak state, with a certain degree of autonomy for Slovakia. This situation proved to be a tactical asset to the Communists in their competition with other political parties, though an unanticipated liability to their own internal cohesion. Finally, whereas in postwar Poland's early years, the Socialists enjoyed significant support in the trade-union movement, the Czechoslovak unions were always under Communist direction.

One politically consequential similarity between the two countries was that both had been liberated from the German occupation by the Soviet army (except, ephemerally, a narrow strip of western Czechoslovakia where the Americans appeared before the Soviets in April 1945). And their own armed forces were directed by defense ministers who, though nominally nonpartisan, were in the pockets of the Communists.

Similarly, the internal-security (police) apparatus in both states was tightly monopolized by the Communists. Although Czechoslovakia did not gain new territory from Germany, as did Poland, it acquired much German property, thanks to the expulsion after the war of about 3 million ethnic Germans. And as in Poland, the Communists took care to control the extensive patronage apparatus accruing from the distribution of the expellees' farms and property to Czech and Slovak beneficiaries, and to depict the Soviet Union as henceforth the only reliable Great Power guarantor of these gains against eventual German revanchism. Finally, in both Poland and Czechoslovakia, the catastrophic events of 1938 and 1939 and then the experience of wartime occupation had sullied the prestige of the interwar parties and systems and the authority of their surviving leaders. Although this change in the moral climate of public life did not automatically give political strength to the Communists, it did erode the stamina and the self-confidence of their domestic competitors.

The government with which President Edvard Beneš returned to Prague in May 1945 had been sculpted in Moscow two months earlier. Its premier was Beneš's wartime ambassador to the Soviet Union, Zdeněk Fierlinger, whose party affiliation was Social Democratic but who soon revealed himself to be a reflexive enthusiast for zombie-like coordination with the Communists and the Soviet Union. His party had two additional ministerial assignments. The non-Marxist National Socialists (who bore no ideological relation to their German namesake and were simply a progressive, bourgeois, anticlerical reform party) and the Czech Populists (Catholic) each had three; the Communists, four. The defense and foreign affairs portfolios were assigned to nonparty professionals, with a Communist being designated deputy minister of foreign affairs so that he could monitor his chief, and the defense minister being a general with warm "brotherhood-of-arms" feelings toward the Soviet armed forces, at whose side he had fought in the war. The semiautonomous regional subgovernment of Slovakia was shared between the Slovak Communists (who had absorbed the local Social Democrats in 1944) and the Slovak Democrats (an amalgam of all the other non-Communist members of the Slovak National Council, which had authorized the uprising of 1944). These five governmental parties—with the Communists appearing in their dual guises of "Czechoslovak" and "Slovak"—constituted themselves as the National Front, proscribed all other interwar and wartime parties as having compromised themselves by treason and collaboration, and permitted no institutionalized opposition.[6] Though the four non-Communist parties had conceded much and the relative Communist weight in the government was to prove heavier than the numerical distribution of portfolios suggested, the National Front of 1945 was still a government of agreed compromise rather than a veiled seizure of power, as was its Polish counterpart.

The initial moderation of the Communists in the distribution of portfolios was matched by a moderation in policy matters. Their calls for alignment with the Soviet Union were not out of line with those of their coalition partners and were, after all, anticipated by Beneš's own treaty with Stalin in December 1943 (see Chapter 2, section 3). In domestic policy, they demanded no radical socioeconomic transformations beyond the collective commitment of all the National Front parties to the nationalization of banks, heavy industry, and large factories and to the distribution of large landed estates to peasants; they expressed respect for private property and Christian values (even tolerating parochial schools) and were quite sparing in their recourse to Marxist-Leninist rhetoric. Nor did they initially abuse their control of the police apparatus and of other instruments of intimidation in anything like the manner of their comrades in other people's democracies. Thus the Czechoslovak Communists nourished a widespread impression, shared by Beneš, that they were different from other Communists, patriots first and foremost, evolutionary reformers, and reliable partners in the national coalition government. Accordingly, all the parties of the coalition set about recruiting mass memberships and developing their organizations. Soon 40 percent of adult Czechoslovaks were members of political parties. Such a high degree of politicization not only was probably unhealthy in a general sense, but also advantaged the Communists specifically—not so much because their membership was the largest of the five parties, but because it was the most tightly organized and firmly disciplined, and hence most capable of infiltrating and eventually manipulating the extrapartisan institutions and mass organizations of public and collective life, such as trade unions, "peasants' commissions," professional bodies, grass-roots "action committees," and local governments.

Nevertheless, for more than two years after the end of World War II, the Czechoslovak Communists refrained from any extravagant flexing of their political muscles. As long as the French and Italian Communist parties were still participating in their respective governments, and while Czechoslovakia was being widely viewed in the West as a test case of the possibility of preserving the wartime Big Three alliance into the postwar era, it made sense for Stalin to try to sustain the viability of the first and third options of Harriman's suggestive triad (see section 1). And his disciples in Prague were powerfully tempted to reach for the enormous prestige that would accrue to them should they achieve a legal, electoral, peaceful conquest of power in their relatively advanced industrialized country.

From this last perspective, the results of the first (and last) free parliamentary elections of the postwar era, on May 26, 1946, were ambiguous.[7] The Communists (Czech and Slovak combined) won 38 percent of the votes throughout the country; the Social Democrats, 13 percent; the National Socialists, 18 percent; the Czech Populists, 16 percent; and the Slovak Democrats, 14 percent. Within semiautonomous Slovakia, the Slovak Democrats won 62 percent; the Communists, 30 percent; and two new minor parties received 4 and 3 percent each. The Communists' performance (40 percent in the Czech lands, 30 percent in Slovakia, and 38 percent in the state as a whole) was impressive, though it seems to have fallen short of their expectations. Their Social Democratic allies were the biggest losers and had to yield the premiership to the Communist Klement Gottwald. Yet as the holder of the balance between the Communists and the non-Marxist parties within the National Front, the Social Democrats became its most courted, solicited, and cajoled member party.

The Communists quickly demonstrated their resiliency, skill, and tenacity. Deeming unacceptable the prospect of losing political control of the Slovak administrative apparatus to the Slovak Democrats, they coaxed the other Czech parties into joining them in passing legislation a month after the elections that sharply limited the autonomy of Slovakia, reducing it to little more than a regional administrative unit of the Prague-centered government. Though it achieved its immediate intentions and was enacted legally and peacefully, this measure reimposed a chronic strain on Czech–Slovak relations that was later to yield a bitter harvest in the "spring year" of 1968. Compounding this source of bitterness was the treason trial during the winter of 1946/47 and the hanging on April 18, 1947, of Monsignor Jozef Tiso, the president of the Axis puppet state of Slovakia. Though they had co-led the 1944 uprising against Tiso, the Slovak Democrats now vainly recommended commutation of his death sentence to life imprisonment, for they grudgingly appreciated that the six years of formal Slovak independence over which he had presided during the war had been symbolically gratifying to the Slovak nation. They were seconded in this clemency recommendation only by the Czech Populists, who balked at hanging a priest, but were overruled by the other Czech parties.

Apart from these tensions between Czechs and Slovaks, the year following the elections of May 1946 was relatively harmonious and constructive. The harvest was good; industrial production leaped; and foreign trade soared, albeit mainly with Czechoslovakia's traditional Western trading partners and scarcely at all with the Soviet Union. The Communists maintained their posture of moderation, and the other parties, which had feared that the Communists might indeed achieve an electoral majority, were relieved and took heart. The National Socialists and the Czech Populists enrolled many new members.

Alas, a combination of foreign and domestic developments brutally crushed this rosy scenario from mid-1947 onward. In May, the French and Italian Communist parties were dropped from their respective national coalition governments; in June, the United States initiated the Marshall Plan for European recovery, and in early July, Stalin insisted that the Czechoslovak government reverse its initially unanimous decision (that is, including the Communists) to participate in it; in September, the Czechoslovak Communists were berated for their relative moderation—their failure to "resolve the question of power"—by the Soviet delegates to the founding session of the Cominform, a session at which the formerly acceptable doctrine of separate national paths to socialism was condemned and the people's democracies were ordered to coordinate their systems and their policies with those of the Soviet model. Domestically, a severe drought blighted the 1947 harvest, leading to food shortages, a black market, blatantly unjust food distribution, and much anger at the Communists, who headed the most relevant ministries of Agriculture, Internal Trade, Social Welfare, and Finance.

The Communists sought to divert this public resentment by radicalizing the situation with such demagogic proposals as a "millionaires' tax," another round of land distribution, more nationalization of enterprises, and forced bank mergers—only to be surprisingly blocked by their coalition partners in the cabinet and the parliament. In November 1947 came an even more unsettling pair of developments as the Communists' hitherto automatic Social Democratic echoers replaced Fierlinger, their fellow-traveling chairman (and former premier), with the reputedly less compliant Bohumil Laušman, and as the Communists' attempt to seize total control of the Slovak subcabinet through an administrative coup was at least parried, though not roundly defeated.

But with Moscow's recent scoldings for alleged softness still ringing in their ears, the Czechoslovak Communists could not afford to yield or retreat, even had they been so inclined (which was not the case). They redoubled their efforts to augment their own power by intimidating and paralyzing the other parties, whose leaders' blunders played into their hands.

Aware that several public-opinion polls projected a sharp decline in its prospective electoral fortunes and unwilling to accept such a humiliation, the Communist party launched a strident campaign that the next parliamentary elections, scheduled for May 1948, be based on a single-ticket list, to be composed in advance by the member parties of the National Front and then ratified by the electorate by plebiscite. The Communist minister of the interior also aborted the investigation of a mysterious provocation entailing the mailing of parcels containing bombs to several non-Communist ministers, and proceeded to purge the police apparatus of its few remaining non-Communist officials. In mid-February 1948, the majority of the non-Communist cabinet ministers, including the Social Democrats, instructed the minister of the interior to reverse this purge. On February 20, after he (backed by his party) deliberately ignored this cabinet resolution, the ministers representing the National Socialist, Czech Populist, and Slovak Democratic parties resigned, hoping (but failing to ensure in advance) that they would be joined by the Social Democrats (which would have entailed the fall of the cabinet) or, failing this, that President Beneš would function as their safety net by refusing to accept their resignations. But they had made no serious plans for complementary or follow-up measures of their own beyond this rather weak act of resignation, with which they precipitated a crisis that immediately overwhelmed them.[8]

Alas for those who resigned, they had miscalculated on virtually all counts—arithmetic, psychological, and political. They numbered only twelve in a cabinet of twenty-six, thus leaving the Gottwald government legally still in office. The Social Democrats retreated from their recent shows of independence and succumbed to Communist pressure and bribes to remain in their ministerial chairs. And the Communists, determined not to permit a replay of the French and Italian scenarios of May 1947, efficiently mobilized the mass organizations that they had earlier penetrated (and improvised new ones, such as "people's militias") to take over the streets, media, work places, public buildings, and utilities, and thus to generate a really prerevolutionary atmosphere. In the face of this pressure (reinforced by the nominally nonpartisan defense minister),

President Beneš, who was constitutionally required to be politically neutral and on whom the resigning ministers had imposed an unreasonable burden by expecting him to spare them the consequences of their political amateurishness through a legal formality, accepted the twelve resignations on February 25, 1948.[9] The Communist premier easily replaced them with tame (and probably suborned) members of their own parties. "The question of power" in Czechoslovakia was thus resolved—not quite in the manner that the Communists had planned, but as the result of their deft exploitation of a simmering crisis that their enemies had brought to a boil.

A trio of interpretive addenda to this narrative is in order. It will be recalled that the Communists had craved a legal, electoral, peaceful conquest of power in this industrially most advanced of the people's democracies. The way in which it occurred in February 1948 was formally quite legal but not electoral; and although the takeover had indeed been bloodless and free from overt violence, the latent threat of force by the Communist-controlled police, army, and "people's militias" was surely decisive. Thus the Communist seizure of power in Czechoslovakia was both constitutional and revolutionary. This brings us to the second point. Many analyses of these events have stressed the presence of Valerian A. Zorin, the Soviet deputy minister of foreign affairs, in Prague during the crisis, as though this proves not only that the Soviets had masterminded the entire Communist performance, but also that only the specter of their possible direct intervention decided the issue.[10] But this interpretation underrates Gottwald and his comrades. Though the Soviets undoubtedly would not have tolerated the expulsion of their protégés from the Czechoslovak government and that country's slippage into the Western orbit, the local Communists proved quite capable of managing the crisis on their own. The third point is that none of the participants in this conflict—the Communists, their foes, the fence-sitters, the president—considered appealing to parliament for a vote of confidence or censure. This was in keeping with the interwar Czechoslovak *pětka* tradition, by which all political decisions and solutions were taken and brokered by party leaders off the floor of the legislature, which was then expected to rubber-stamp them.[11]

The Communists' mopping up after their breakthrough of February 20 to 25, 1948, was efficient. Anti-Communist newspapers and periodicals were closed and non-Communist ones purged during the next few days, followed shortly by universities, professional bodies, sports clubs, the publishing industry, and the civil and military services. In April came the nationalization of all enterprises employing more than fifty workers and all engaged in foreign or wholesale trade, as well as breweries, bakeries, and dairies—thus ending the embarrassing spectacle of establishments left in private hands after the initial nationalizations of 1945 prospering to the shame of the less productive state-owned ones. The nationalization of industry and commerce was paralleled by a necessarily slower but nevertheless relentless drive to socialize agriculture, a drive that by 1960 had collectivized over 90 percent of the land. A typically Stalinist five-year plan was launched on January 1, 1949, with the announced purpose of "eliminating all traces of capitalism" from Czechoslovakia's economy. It deemphasized the light industries that had traditionally produced goods for export to the West and stressed heavy industry and coordination with the economies of the Soviet Union and the people's democracies. Thus whereas before World War II, only one-tenth of Czechoslovakia's foreign trade had been with its eastern neighbors, by 1954 it was four-fifths. In 1951, the five-year plan was revised to reinforce the industrialization of Slovakia.

On the narrower political front, the Communists' brisk mopping up after February 1948 included the extension of their purge to Slovakia in March, the reduction of all other parties into bogus window dressing in April,[12] the adoption of the new Communist-tailored constitution accompanied by single-list ratification elections in May, the enforced merger of the self-castrated Social Democratic party with the Communist party in June, and the introduction of the ominous notion of "subjective guilt" into judicial procedure, together with the unleashing of police terror, in October.

The Communists' shift from cooperating with the Roman Catholic church to seeking to subdue it began promptly that same autumn of 1948—much earlier than in Poland. The assault was also more successful than in Poland in the sense that—after several humiliating trials of churchmen, the banishment of Archbishop Josef Beran of Prague to a provincial monastery,[13] extensive confiscations of ecclesiastical properties, and other devastating fiscal and administrative pressures (mixed with rewards for collaborating priests)—the vast majority of the lower clergy and nine of the thirteen functioning bishops yielded by swearing oaths of loyalty to the people's democratic regime (not simply to the Czechoslovak republic) by the spring of 1951. And although the fundamental commitment of the masses to Roman Catholicism appeared to remain intact, especially among the Slovaks, a regime-sponsored movement of so-called patriotic priests purporting to represent a synthesis of

Christianity and Marxism resonated more deeply in Czechoslovakia than in Poland.

Thus the Czechoslovak Communists, who had seemed to be the "softest" in East Central Europe—with their initial postwar stance toward democratic institutions, competing parties, and the West—were to emerge during the 1950s as the area's arguably most Stalinist party—with their combination of repressiveness, rigidity, xenophobia, and ferocity of internal "anti-Titoist" purges (see Chapter 4, section 3).

4

In contrast to both Poland (where Stalin flatly deemed his claims to Soviet hegemonic control to be nonnegotiable) and Czechoslovakia (where for an extended interval he attenuated his other imperatives in order to maintain a bridge to the West), Soviet postwar policy in Hungary was initially rather fitful, as though Stalin was not quite certain whether that country could be permanently subsumed into the Soviet and Communist political orbit. Indeed, had Regent Miklós Horthy succeeded in his clumsy and, in the event, abortive maneuvers to negotiate an armistice and switch sides in the autumn of 1944, Stalin would have been content to leave him in office rather than work with the small and historically rather discredited Communist party. And in mid-October 1944, Stalin recommended to Churchill that British and American forces open a new front on the northern Adriatic littoral and press into Central Europe from there, thus suggesting a readiness to share the liberation of Hungary with them. Furthermore, in the provisional Hungarian government that the Soviets finally unveiled on December 23, 1944, the Communist representation was still minuscule compared with what it was in the Soviet-sponsored Polish and Czechoslovak governments. And this provisional government's program was strikingly moderate, its only socioeconomic innovation being the reform of a landed estate system that was more or less universally acknowledged to be structurally primitive, unproductive, and notoriously inegalitarian. Even the behavior of the Soviet military administration in Hungary between the autumn of 1944 and the summer of 1945, characterized by indiscriminate requisitioning, dismantling, and removal of stocks, assets, equipment, and facilities, hints that the Soviets did not expect to stay long.

This hesitancy and ambivalence of the Soviet and Hungarian Communists toward taking power persisted for some time after the end of the war. That it was indeed ambivalence rather than fine-tuned subtlety is indicated by its improvisational and spasmodic quality. On the one hand, the Soviet administrators of the Allied Control Commission insisted on having enormously extensive executive, monitoring, and interdicting jurisdictions. On the other hand, they frequently forbore to exercise them, as when they let József Cardinal Mindszenty, the newly appointed (September 1945) Roman Catholic primate of Hungary, condemn the Marxist parties and the provisional government's socioeconomic legislation (including the land reform), express sympathy for the vanquished Horthyites, and even compare the country's recent liberation by the Soviet armies with its thirteenth-century ravaging by the Tatars. Similarly enigmatic was a remark made by the Soviet chairman of the Allied Control Commission—no less a figure than Politburo member Marshal Kliment Ye. Voroshilov—to the effect that the Soviet Union wished to rely on the catch-all Smallholder party as its fulcrum in Hungary. The Soviets and the local Communists also permitted the revolutionary "national committees" and "people's courts" that they had initially sponsored and controlled in the winter of 1944/45 to be emasculated and absorbed into Hungary's traditional municipal and ministerial structures. By June 1945, 80 percent of the Horthyite bureaucrats were still or again in office, and no effective purge of that bureaucracy was to occur for another two years. The Soviet and local Communists were also deferential toward non-Communist political parties, which were treated far more gingerly than their objective strength warranted and than they would have been had Stalin by then decided to absorb Hungary into his orbit. When the Social Democrats objected to a Communist plan to restructure the trade unions along "industrial" rather than "craft" lines, the Communists yielded in February 1945 and subsequently tolerated a substantial reassertion of Social Democratic influence in the union movement. And on such tangible workers' issues as strikes, wage increases, the functions of "factory committees," and nationalization, the Communists took a less friendly and less rhetorically revolutionary stance than did the Social Democrats. On balance, the Communists' political program in 1944 and 1945 was remarkably self-effacing and self-abnegating. It called for the rule of law, free culture, free intellectual inquiry, free political dialogue, a free press, free enterprise, and free elections.[14]

But when such free elections were held, the results indicated that this political forbearance had earned the Communists neither credit nor popularity. The municipal elections of October 7, 1945, in Budapest—the country's capital and its leading industrial (working-class) and intellectual center—gave to a Communist–Social Democratic list only 42.8 percent of the votes, with 50.5 percent going to the Smallholders, 2 percent to the National Peasants, and 3.8 and 0.9 percent, respectively, to two bourgeois liberal parties styled the Citizen Democrats and the Radicals. Thereupon, the Social Democrats slipped from their Communist electoral albatross and ran independently in the national parliamentary elections of November 4, 1945—the freest ever held in Hungary. This time, the Smallholders won 57 percent of the votes; the Social Democrats, 17.4 percent; the Communists, 17 percent; the National Peasants, 6.9 percent; the Citizen Democrats, 1.6 percent; and the Radicals, 0.1 percent. The potential damage to the Communists was limited by a preelection agreement of the parties composing the National Independence Front to maintain their coalition government intact no matter what the electoral outcome, an agreement that now prevented the Smallholders from capitalizing on their absolute majority to form a one-party government. Nevertheless, the elections' results were sufficiently disconcerting to prompt a Communist reassessment, both in Budapest and in Moscow, of the past year's self-restraint, which had entailed so much fitful and inconsistent behavior. Mátyás Rákosi, the leader of the Hungarian Communists, was summoned to Moscow to review his party's performance.

The outcome of that reassessment was, however, still quite modest in comparison with the political orientation then prevailing in Poland, Yugoslavia, Romania, Bulgaria, and Albania. The Hungarian Communists still did not decide to seize political power. Rather, like their Czechoslovak comrades, they retained their authentic coalition strategy but henceforth prosecuted it less waywardly and with more sustained, albeit still only incremental, pressure on their partners. They insured themselves against the haunting specter of a possible special suballiance within the National Independence Front of the Smallholders and the Social Democrats, two parties that had drawn close during the war. When Zoltán Tildy, the Smallholder leader, was elevated to the presidency of the newly proclaimed republic on February 1, 1946, the Communists maneuvered to ensure that he would be succeeded as the leader of that party and as prime minister of the government by the more malleable Ferenc Nagy rather than by his tougher competitor, Dezső Sulyok. The next month, they successfully pressured Nagy and Tildy into purging Sulyok (together with twenty-one other Smallholder parliamentary deputies) from his party, ostensibly because he was spoiling Hungary's relations with supposedly fraternal Czechoslovakia by his protests over the latter's hard treatment of its remaining Magyar minority, but in fact because he was deemed recalcitrant to Communist leverage, and hence "reactionary." Furthermore, in Nagy's new government, the Ministry of the Interior, with its critically important control of the police, surveillance, and security apparatus, was for the first time assigned to the Communists—never thereafter to be relinquished. Interestingly, the Hungarian Communists had initially reconciled themselves to this portfolio going to the Smallholders, but that arrangement was overruled by the Soviets.[15] And when Marshal Voroshilov subsequently demanded the dissolution of some religio-social and youth organizations, Nagy again complied, despite considerable restlessness among members of his party. An important reason for the obsequious behavior of the leaders of the majority Smallholders toward the Soviets and the Communists during these months is that they hoped (vainly, as events were to show) thereby to persuade Moscow to lighten Hungary's reparations burden[16] and to favor its claims for the retrocession of at least part of Transylvania from Romania at the forthcoming European peace conference, scheduled to open in July 1946. Conversely, the still relative Communist moderation of this period was partly predicated on a reluctance to provoke the British and Americans in advance of that peace conference. Indeed, Rákosi even joined Nagy on a ministerial delegation that visited Washington and London in May 1946 to solicit support for Hungary's case. Stalin permitted no other country that had been liberated by his army to make such an open political overture to the West.

Once the peace treaty with the Allies had been signed on February 10, 1947, the confrontation in Hungary sharpened. The Smallholders, disappointed in their hopes of obtaining Soviet support through propitiation, now stiffened and rejected a Communist demand that they repudiate their executive secretary, Béla Kovács (as they had purged Sulyok the previous year), whereupon the Soviet authorities (not the Hungarian police) simply arrested him on February 25, in violation of formal Hungarian sovereignty. The Americans and the British protested, to no avail. In May 1947, the Soviets announced that Kovács's interrogation had incriminated Prime Minister Nagy as participating in a conspiracy against the republic. Nagy, then on vacation in Switzerland and fearing arrest were he to return home, resigned in exchange for the release of his four-year-old son and remained in exile. In July, Hungary was obliged by the Soviet Union to join its neighbors in declining to participate in the American-sponsored Marshall Plan for European economic recovery. Instead, on August 1, a three-year plan providing for a state-directed economy (but not yet for steep Stalin-type capital investments, which awaited the five-year plan of 1950) went into effect. By this time, also, the civil bureaucracy had been purged and politicized to the point where the staffs of governmental agencies and offices that were headed by non-Communist ministers were no longer responsive to their nominal chiefs without Communist assent.

Yet in the parliamentary elections of August 31, 1947—which were less free than those of two years earlier due to considerable intimidation and fraud, but nevertheless still competitive and more free than those in Hungary's Balkan neighbors—the Communists received only 22.3 percent of the votes; the Social Democrats, 14.9 percent; the now lacerated Smallholders, 15.4 percent; the National Peasants, 8.3 percent; the Citizen Democrats, 1 percent; and the Radicals, 1.7 percent. Four new parties ran in declared opposition to the National Independence Front, but since they, in effect, drew votes away from the Smallholders, their appearance was quite welcome to the Communists: the conservative, nationalist Independence party polled 13.4 percent; the more progressive Democratic Populists (Catholic), 16.4 percent; a list headed by the priest-politician István Balogh, a former Smallholder and an accomplished intriguer, 5.2 percent; and the Christian Women's League, 1.4 percent. At a minimum, these elections indicated that the public had not yet been cowed into helpless acquiescence.

But the Communists, though embarrassed, were not stopped by their poor electoral performance. On November 15, 1947, the new parliament was pressured into delegating its powers to the government and proroguing itself for over a year. That same month, the Independence party was dissolved and its leader forced to flee abroad. Slightly more than a year later, a similar fate was inflicted on the Democratic Populists. Meanwhile, in March 1948, the Social Democrats were compelled to withdraw from the Socialist International and, in June, to purge, dissolve, and merge themselves with the Communists into the latter's formally renamed Hungarian Workers party. Then on July 30, 1948, Tildy, a Smallholder, was obliged to resign the presidency of the republic in favor of the chairman of the Workers party. A third cycle of parliamentary elections was held on May 15, 1949. This time there was to be no pretense of democratic procedures. The secret ballot was reduced to a farce, and a single "Government List," drafted by the Communists, was proclaimed to have been endorsed by 95.6 percent of the voters. The new parliament promptly adopted a Soviet-emulating constitution under which Hungary was officially designated a people's democracy. With exquisite irony, it went into effect on August 20, 1949—the traditional national feast day of Hungary's patron saint and first king, Stephen (István).

As the process of bending, breaking, and coordinating the secular political institutions was approaching its denouement, the Communists turned on Hungary's ecclesiastical bodies. Like postwar Poland, Hungary was by now a well-nigh homogeneous country from an ethnonational perspective; but unlike Poland, it was religiously pluralistic, with approximately two-thirds of the population Roman Catholic; one-quarter, Calvinist; and very small fractions, Lutheran and Jewish.[17] As the Catholic church was historically identified with the imported Habsburg dynasty and as its prelates were often Magyarized former Slovaks and Schwabs, the Calvinists, who were concentrated in the eastern part of the country, tended to view themselves as purer Magyars than the Catholics. This attitudinal distance between the two denominations, though not wide enough to be deemed a profound alienation, provided an opening for anticlerical political leverage. Thus whereas Cardinal Mindszenty rather anachronistically called for a Habsburg restoration and emphatically opposed the postwar land reform, the Calvinist ministry endorsed the republic and welcomed the land reform as both necessary and overdue.

In the realm of culture and pedagogy, both these churches entered the postwar era still enjoying very extensive institutionalized public authority. Two-thirds of all primary schools were confessional, and religion was obligatorily taught even in the state primary schools. One-half of the boys' and four-fifths of the girls' secondary schools were Catholic. In addition, the Catholic church had been very wealthy until the 1945 land reform.

The postwar governments initially adopted a rather cautious attitude toward the churches, favoring them with priority allocations for the reconstruction of their wrecked or damaged places of worship and with generous annual budgetary grants. But the deeply conservative, obstinate, vehement, and courageous Cardinal Mindszenty was negative and combative from the start—in contrast to Poland's socioeconomically more progressive and politically more sophisticated Primate Wyszyński. When in 1947 the state schools introduced new textbooks for secular subjects, which the Calvinist schools also adopted, Mindszenty forbade their use in Catholic schools. When in June 1948 the two confessional primary-school systems were nationalized, albeit with the retention of religious instruction, he excommunicated the Catholic governmental officers who supported this law and instructed the teaching priests and nuns to withdraw from the schools. Over 4,500 did so initially, but later returned because they were obviously needed to teach the ongoing religious lessons that most parents ostentatiously requested even though they were now optional.

As in Czechoslovakia, so in Hungary, 1948 was a year of belated and therefore accelerated Communist monopolization of power. Precisely therefore, Mindszenty—whose unbending rejection of virtually all post-war socioeconomic and political developments had seemed quixotic in 1945—had become something of a vindicated hero in popular opinion, and the Communists deemed it necessary to discredit him publicly. Arrested the day after Christmas 1948 and tried in early February 1949 on charges of currency speculation, espionage on behalf of imperialists, and treason against the Hungarian republic, he proved to be a surprisingly abject defendant, pleading guilty to most charges and confessing that his previous attitude had been wrong.[18] But the regime's potential domestic propaganda success was outweighed by the international revulsion provoked by the sinister ambience of this trial. Mindszenty was sentenced to life imprisonment (ameliorated to house arrest in 1955). Then most monastic orders were dissolved, and the monks and nuns evicted from their cloisters. (In Czechoslovakia, too, the religious orders were suppressed, whereas in Poland they functioned freely.) On August 30, 1950, the Catholic church yielded significant political concessions in return for the restoration of a mere 8 (out of more than 3,000!) of its nationalized schools and the stabilization of the state's financial subsidies to its clergy. It promised to support the five-year economic plan, the collectivization of agriculture, and the Soviet-sponsored Stockholm Peace Appeal; to disavow subversion; and to pledge allegiance to the people's republic. Nevertheless, some arrests and trials of clergy continued. In 1952, the Calvinist ministry was also purged and obliged to yield its last remaining high schools to the state.

Meanwhile, as in the other people's democracies, a severe inner crisis wracked the Hungarian Communists precisely at the culmination of their victory over the country's other parties and political forces. As this crisis was an integral aspect of the rift between Tito and Stalin, which wrenched the Communist movement throughout East Central Europe in the years after 1948, it will be analyzed in Chapter 4. Suffice it to note here that in Hungary, the crisis turned on the issue of the "correct" or "erroneous" quality of the earlier phase of Communist restraint and moderation and of responsibility for that "line." That earlier Communist and Soviet stance in Hungary—which was not merely restrained and moderate, but also contradictory and even capricious—has also been viewed in this book as a puzzle to which several possible solutions have been tangentially suggested. The most probable is that the stance was an outgrowth of Stalin's general uncertainty about his postwar relations with the United States and Great Britain. On the one hand, he hoped to maintain the wartime alliance; on the other hand, he feared that if and when it came to an end, the Americans would press him sharply. Specifically, he was bracing himself for the possibility that he might have to release Hungary from his political orbit in return for a free hand in Poland, which had a far higher priority for him. When the spirit of the wartime alliance finally did give way to the Cold War in 1947, Stalin was both rendered anxious and relieved that the Americans sought only to "contain," not to "roll back," Soviet power in East Central Europe. Thus he was not, after all, subjected to the anticipated pressure to trade Hungary for Poland and could unleash his Communist protégés in the Danubian state for their belated rapid seizure of power after 1947.[19]

5

In Yugoslavia, the Communist seizure and even consolidation of power had occurred during World War II, which was there a civil war and a revolutionary war as well as a war of national liberation. Indeed, postwar Yugoslavia leapfrogged over the phase of multiparty coalition, which in the other people's democracies had at first a substantive reality of variable duration and then a nominal continuity. The Yugoslav People's Front was instead defined from the start as a "bloc of Communists and nonparty sympathizers"[20] and used as a bludgeon to smash other political parties and loyalties. Hence in the elections held on November 11, 1945, for a two-chamber constituent and legislative assembly, the choice was simply between endorsing the Communist People's Front or visibly putting one's ballot into a residual "opposition urn." Predictably, the results were as follows: Federal Chamber—88.7 percent of the eligible voters cast ballots, of whom 90.5 percent endorsed the People's Front and 9.5 percent opposed it; Chamber of the Nationalities—88.4 percent of the eligible electorate voted, of whom 88.7 percent endorsed and 11.3 percent opposed the People's Front. Formally, symbolically, and substantively, the two-chamber assembly, the elections for it, and the republican, federal constitution that it adopted on January 31, 1946, imitated the Soviet pattern much more closely than did analogous institutions and procedures in other people's democracies at this early date. Yet it would be erroneous to impute these lopsided election results purely to intimidation and to overlook the deep reservoir of authentic popularity, prestige, and legitimacy that Tito, his Partisan movement, and his Communist party had amassed during the war. It would also be a mistake to underestimate the Titoist system's extensive generosity toward the distinctive cultural, linguistic, and educational aspirations, but not the centrifugal political orientations, of Yugoslavia's several component ethnonations. For example, the decision to again designate the federation's constituent republics by their ethnohistorical names, which the interwar royal government had abolished and suppressed in 1929, was important and gratifying.[21]

The symbolic coda to this essentially wartime Communist seizure of power was the execution on July 17, 1946, of Tito's only comparable competitor of those years, the Četnik leader General Draža Mihajlović, who had been hunted down in Bosnia in March. As this execution aroused some murmuring among Serbs, Tito's regime evened the domestic ethnonational score, as it were, on October 11, 1946, by sentencing Alojzije Stepinac, the Croatian Roman Catholic archbishop of Zagreb, to sixteen years of imprisonment at hard labor for war crimes and collaboration with the Axis occupiers. Stepinac's wartime behavior had been ambiguous and subject to contrasting interpretations. Because he had not publicly intervened to halt the forcible conversion of Orthodox Serbs to Catholicism (and the massacres of those who refused) by Ustaša bands, he had become a symbol of genocidal enmity to the Serbs. But to Croats, even those who had distanced themselves from the Ustaša, his sentence seemed an act of ethnonational spite. As the Tito government was then pursuing a hyper-radical, ultrarevolutionary line in both its domestic and its foreign policies, the trials of Mihajlović and Stepinac were orchestrated in order to implicate and discredit the West.[22]

The Communists' capture of political power and of the Yugoslav state having been coterminous with the war, their postwar revolution was immediately prosecuted in the economic and sociocultural realms. This entailed not only centralized direction of the economy as well as heavy investment and extensive nationalization in the industrial, commercial, and credit sectors—a formidable challenge for a country that was underdeveloped to begin with and had been devastated by the war—but also a fierce drive to transform the so-called petit-bourgeois outlook of the peasantry through the collectivization of its land—a drive that the peasants resisted as stubbornly as the regime prosecuted. Calling for the reinvestment of between one-quarter and one-third of the national income, the regime's economic plan was clearly overambitious and was plagued by low productivity and a shortage of skilled labor and personnel; it was soon in serious trouble.

Nevertheless, precisely to justify their claims to ideological purity and to refute the Soviet accusations of "revisionism" and "capitulation to capitalism" during the incubation of the Tito–Stalin rift in 1947 and 1948, the Yugoslav Communist leaders initially redoubled their investment, development, and collectivization drives in the late 1940s, even though the Soviet Union and the other people's democracies were compounding Yugoslavia's economic crisis by reneging on their trade commitments to it. By the turn of the decade, a greater proportion of industry, commerce, and agriculture had been "socialized" in Yugoslavia than in any other people's democracy. But the entire economy was exhausted and nearing collapse. Ominously, peasant riots erupted in 1950 in some of the regions that had first backed the Partisans during the war. But though economically irrational, the ideological superorthodoxy of those years performed the political function of enabling the Yugoslav Communists to survive Stalin's onslaught with their revolutionary self-respect and their organizational self-confidence intact. Having served this purpose during a period of acute crisis, it could be relaxed and succeeded by greater economic realism after 1952. Trade was then expanded with the West; investment goals were moderated; and peasants were permitted to withdraw themselves and their land from collective farms (decree of March 30, 1953). Only 1,000 out of almost 7,000 collective farms survived, and by 1954 over 80 percent of the agricultural land was back in private ownership. Other structural, socioeconomic, and political changes soon followed as Yugoslavia transformed itself into a unique type of Communist-controlled but not Soviet-modeled society—a process to be analyzed in later chapters.

6

It was widely expected that the Romanian royal coup and attendant switch from the Axis to the Allies, executed on August 23 to 25, 1944, would be vigorously contested by the Germans. Hence it was deemed appropriate—apparently also by the Soviets—that the first postcoup Romanian government consist primarily of military men, with the relatively ornamental addition of four civilian leaders of the National-Peasant, Liberal, Social Democratic, and Communist parties as ministers without portfolio. The only civilian politician assigned a portfolio, that of Justice, was the hitherto "local underground" Communist Lucreţiu Pătrăşcanu, who also doubled as one of the four ministers of state.[23] Two months earlier, in June, these four parties, formally still illegal, had formed themselves into an ad hoc coalition to press for Romania's leaving Hitler's war—a goal to which they had now been anticipated by the king's action. The prime minister of the new government was General Constantin Sănătescu, marshal of the royal court and the link between King Michael and the army in the recent coup. One of his first acts was to restore the constitution of 1923, which King Carol had suspended in 1938, thereby relegalizing political parties and political life.

By early November 1944, it was clear that the *Wehrmacht*, though indeed fighting tenaciously to hold Hungary, was no longer capable of a thrust into Romania. Hence the addition of civilians to the Sănătescu government was now deemed feasible. Though the general remained prime minister and became acting war minister, he was henceforth the cabinet's only military member, all other portfolios going either to nonparty civilian specialists or to members of the four coalition parties. Interestingly, the two new Communists holding portfolios, Gheorghe Gheorghiu-Dej (Transportation and Communications) and Vlădescu Răkoasa (Minorities) were, like Pătrăşcanu (who retained Justice), "locals." The top "Muscovite" Communists of that time, freshly returned from Soviet asylum—Emil Bodnăraş, Vasile Luca, and Ana Pauker—did not yet assume governmental posts and instead devoted themselves to rebuilding the weak and depleted party apparatus and its paramilitary organs. Pauker would later concede that at this time the Communists, emerging from two decades underground, numbered fewer than 1,000 members.

A genuine coalition cabinet, the second Sănătescu government lasted only a month because the Communists objected to its minister of the interior (a hostile National-Peasant) and the Soviets charged it with being dilatory in fulfilling Romania's armistice obligations. To signal his displeasure, Stalin delayed transferring the administration of northern Transylvania to Romanian responsibility while a small-scale guerrilla war raged there between Hungarian and Romanian bands. Taking the hint, King Michael replaced Sănătescu on December 6 with another general, Nicolae Rădescu, who was presumed to have credit with Moscow thanks to having been interned during the war for his anti-German attitude. Rădescu took over the interior portfolio (with a "local" Communist as undersecretary) as well as the premiership and replaced the Communist minister for minorities with a nonparty specialist, but otherwise made no politically significant changes in the cabinet that he inherited from Sănătescu.

The initial expectation that the Rădescu government would prove stable was undermined early in the new year when the "local" Communist Gheorghiu-Dej and the recently returned "Muscovite" Pauker visited Moscow and were either directed or permitted to bid for power. The instrument through which they did this was the so-called National Democratic Front, formed in October 1944 and composed of the Communist party, the Social Democratic party, the Plowmen's Front (an independent radical peasant party in the 1930s that had been heavily infiltrated by the Communists and by now was virtually their rural arm), the Union of Patriots (a Communist front among the intelligentsia and the professional and business classes), and the trade unions. In mid-January 1945, the National Democratic Front began to agitate for radical land reform, a purge of "reactionaries" and "war criminals," and the "democratization" of the army, and to present itself as alone capable of persuading the Soviet Union to transfer northern Transylvania to Romanian control. Its campaign was ominously seconded by a resumption of Soviet allegations that the current Romanian government, like its predecessor, was lax in discharging its armistice obligations. Marked by increasingly violent street demonstrations, strikes, and land seizures that, in turn, were lubricated by the Communist undersecretary of the interior's sabotaging the instructions of his nominal chief, General Rădescu, and facilitated by the absence of most of the army from the country at the front (see Chapter 2, section 6), this bid for power by the National Democratic Front (actually by the Communists) provoked the exasperated Rădescu into an intemperate radio speech on February 24 in which he denounced some Communist leaders by name as "venal foreign beasts." (As their names indicate, several of these leaders were indeed of ethnically Jewish, Magyar, or Ukrainian descent.) Three days later, Soviet Deputy Foreign Minister Andrei Ya. Vyshinsky flew into Bucharest, where Soviet troops simultaneously occupied the army headquarters and other government buildings, and browbeat King Michael into dismissing Rădescu. The monarch balked for another week at meeting Vyshinsky's supplementary demand that Petru Groza, the leader of the Plowmen's Front, be designated the new premier. But on March 6, 1945, this, too, was done. Groza's new government was nominally a coalition in which Communists headed only three ministries (but had undersecretaries in several others), while two dissident Liberals from outside the National Democratic Front were assigned the prestigious portfolios of Foreign Affairs and Finance, but it was altogether subservient to the Soviet Union and responsive to Communist pressure.[24] Its immediate reward was Stalin's permission, given on March 9, to extend the Romanian administration to northern Transylvania. This enabled the Romanian Communists to capitalize on nationalist sentiment, while the Hungarian Communists were reduced in later years to arguing (unconvincingly) that Stalin's verdict would have fallen the other way had Hungary been less laggard in the pace of its "democratization." It should be noted, finally, that this early and quite decisive Romanian crisis of January to March 1945 bracketed the Big Three's Yalta Conference, held from February 4 to 11, 1945, and seriously disturbed American policy makers about Soviet intentions. Thus, ironically, just as Romania had in 1940 been the first bone of contention in the alliance of 1939 between Hitler and Stalin, so it now occasioned the first open rift between Roosevelt and Stalin as the Americans flatly refused to recognize the Soviet-imposed Groza government. The British followed the American lead only reluctantly, for Churchill believed that he owed Stalin a quid pro quo for having turned a blind eye to the suppression of the ELAS–Communist revolt against the royal Greek government three months earlier.

Conventionally depicted as merely an opportunist, a puppet, and a buffoon, Groza, albeit eccentric, was a more serious political figure than this caricature suggests. He was a prosperous Transylvanian landowner, industrialist, and lawyer; had been educated in Budapest and was fluent in Magyar as well as Romanian; occasionally served as a minister in Romanian cabinets since 1920; and had founded the Plowmen's Front in 1933. Groza was now committed to three major political propositions: (1) carrying out land reform for the benefit of the rural proletariat and smallholders; (2) burying the traditional feud with Hungary as pointless and mutually harmful; and (3) accepting that geopolitical factors required Romanian accommodation to Soviet policy, not for sentimental or ideological reasons (which were operative for the true

Communists) but simply as rational *raison d'état*. All of them were controversial, two were unpopular, none was unreasonable, and, in the long run, all were abortive.

Accordingly, a land-reform program, which Rădescu had wanted to postpone until after war's end and demobilization, was decreed by the Groza government on March 20, 1945, two weeks after its tumultuous installation. The brusque and hectic atmosphere in which the redistribution was effected, the political slanting that characterized its official statistical accounting, and the fact that it was presently superseded by enforced collectivization render any assessment of its effectiveness highly problematic. All that can be said with assurance is that its magnitude, in terms of both acreage distributed and peasant recipients, was less than that of the big "bourgeois" Romanian land reform of 1918 to 1921 and that it, like the earlier reform, failed to transform the Romanian village into a society of prosperous smallholders. Neither of these land reforms, nor both of them cumulatively, solved the problems of rural overpopulation, low agricultural productivity, strip farming, and accelerating fragmentation of holdings. But each did neutralize (not satisfy) the peasantry at a critical political moment in Romanian history. As for Groza's wish for reconciliation with Hungary, his government initiated an authentic effort to end the traditional reciprocal jingoisms by granting generous ethnocultural minority rights, funding a Magyar university, and establishing an autonomous Magyar administrative region in northern Transylvania. Alas, these potentially productive innovations did not survive the revival of militant Romanian nationalism in the 1960s after Groza had passed from the political scene. (He died in 1958.) And, finally, Groza's commitment to Romania's coordination with the Soviet Union as a dispassionate act of state policy was also eventually superseded, first by Stalin's quite profound structural and ideological requirements in and from the people's democracies (see section 1) and second, ironically, by the Romanian Communists' sponsorship of anti-Soviet nationalism since the 1960s.

Meanwhile, in the 1940s, Groza's government consolidated itself in roughly three phases. First, in the autumn of 1945, it faced down an effort by King Michael to exploit the refusal of Great Britain and the United States to recognize it as leverage to replace it with a more acceptable alternative. Assured of Soviet support, Groza simply ignored the royal request to resign, whereupon the king withdrew to his country estate and declined to sign state documents. On November 8, truckloads of the Communist party's militia and troops of a special Soviet-trained army division fired on and broke up a Bucharest crowd celebrating the king's birthday. Yet, oddly, neither the royal court nor the Groza government openly denounced each other, both denying that this anomalous situation (including the piquant spectacle of a king on strike in a Communist-dominated country) amounted to a rupture. Indeed, at the Moscow Conference of the Big Three's foreign ministers, held from December 16 to 26, 1945, a face-saving formula was worked out, whereby a pair of authentic (not fellow-traveling) National-Peasant and Liberal ministers was to be added to the government, which would thereupon receive British and American recognition (duly extended on February 4 and 5, 1946) and then conduct free elections open to all democratic parties. In retrospect, it appears that the American and British negotiators at the Moscow Conference were naïve, hypocritical, or remiss in agreeing to extend recognition before the promised elections, which, when held belatedly on November 19, 1946, were (predictably) quite other than free and open, as an old Romanian tradition of ballot falsification was compounded by newer Communist techniques of intimidation.

The year-long delay before holding the elections was required by the National Democratic Front's (that is, the Communist party's) need to sap the prospective opposition. It delineates the second phase of the Groza government's consolidation. The two new ministers, supposedly appointed to broaden the cabinet, were left without portfolios and simply ignored by their colleagues, and their parties were harassed. In March 1946, the Social Democrats split over the issue of running on a joint list with the Communists, the demoralized majority agreeing to do so. In April, Tătărescu's dissident Liberals likewise jumped on this bandwagon. During the summer, various decrees and laws were issued abolishing the senate, authorizing preelection censorship and discrimination, and otherwise fine-tuning and manipulating the electoral process to ensure the victory of the National Democratic Front, now consisting of the Communists, the purged Social Democrats, the Plowmen's Front, the National Populists (the renamed successor to the Union of Patriots), and the dissident Liberals. In the final count, this bloc was declared to have received 4,766,000 votes, entitling it to 348 parliamentary seats, while the opposition National-Peasants and Liberals were relegated, respectively, to 880,000 votes for 32 seats and 289,000 votes for 3 seats. About 1 million other ballots were distributed among nominally unaffiliated but actually tame Magyar, Democratic, and splinter parties, accounting for the unicameral legislature's 31 remaining seats.

The two authentic opposition ministers who had been added to the cabinet at the beginning of the year now resigned, and their National-Peasant and Liberal party colleagues likewise refused to take their parliamentary seats in protest against the elections' irregularities. The American and British governments piously denounced the process as violating the commitments of the Moscow Conference, but did not withdraw their diplomatic recognition and went on to sign the peace treaty with Groza's freshly purged Romanian government on February 10, 1947.

The years 1947 and 1948 marked the third phase in the government's consolidation, in which the Communists exposed their power ever more openly while retaining Groza as head (until June 2, 1952). The spring of 1947 saw a wave of nocturnal, unexplained, and thus deliberately terrifying arrests of opponents. In July, a number of National-Peasant leaders were apprehended while attempting to flee the country, whereupon that party was outlawed. In October and November, its two top leaders, Iuliu Maniu and Ion Mihalache, were subjected to show trials on charges—not unfounded, albeit vastly exaggerated—of conspiracy with British and American intelligence agents and were sentenced to solitary imprisonment for life. By November, Tătărescu and his fellow dissident Liberal minister had exhausted their usefulness and were dropped from the cabinet, to be succeeded by "Muscovite" Communists who for the first time openly assumed ministerial portfolios: Pauker at Foreign Affairs, Luca at Finance, and Bodnăraş at Defense. On the year's penultimate day, December 30, the still anomalously reigning King Michael was obliged to abdicate shortly after his return from attending the wedding of Britain's Princess Elizabeth and his announcement of his engagement to Princess Anne of Bourbon-Parma. Romania now officially became a people's democracy and received a corresponding new constitution on April 13, 1948 (to be followed by two more constitutions on September 24, 1952 and August 21, 1965, of which the latter was "socialist" rather than "people's democratic").

Politically, 1948 was a year of mopping up and of internal Communist purge. In February, the rump Social Democrats were forced to merge with the Communist party (formally styled the Romanian Workers party between 1948 and 1965), and the dissident National-Peasants with the Plowmen's Front. These two "united" parties, together with the National Populist party (which a year later declared itself sociohistorically superfluous and dissolved) and the tame Magyar party, thereupon presented themselves as the Democratic Popular Front in elections held on March 28 for a parliament to ratify the people's democratic constitution. Predictably, this bloc received 405 seats, ostensibly for 6,959,000 votes. The surviving Liberals and Democrats, running as a spurious and in fact collaborating opposition, were respectively allocated 7 and 2 seats, supposedly for 213,500 and 51,000 votes. Yet this series of affirmations of Communist power did not signal tranquillity. The year was marked by many silent arrests of past and potential opponents as well as by a long series of show trials of yet more National-Peasants, Liberals, military personnel, industrialists, students, clergy, and Zionists. It also saw the purge of the veteran Justice Minister (since the royal coup of August 1944) Pătrăşcanu and other "local" Communists by the high-riding "Muscovites."[25]

Structurally, 1948 marked (1) the residual nationalization of mining, banking, insurance, transportation, and the principal industries, (2) the extension into wider fields of joint Soviet–Romanian corporations (initiated in 1945), controlling important sectors of the Romanian economy under direct Soviet administration,[26] (3) the organization of state farms and machine tractor stations, signaling the next year's launching of the drive to collectivize agriculture, (4) the beginning of "state planning" for the entire economy, and (5) the standardization of the armed forces to the Soviet model.

Finally, 1948 was also the year in which Romania became the symbolic administrative center of European Communism, as the headquarters of the Cominform were transferred from Belgrade to Bucharest in the aftermath of the rift between Tito and Stalin.

It remains now to make a comment on the British and American—especially the American—role in the political developments of 1944 to 1948. The American stance, of repeatedly calling for a Romanian government that would be freely elected and truly representative as well as friendly to the Soviet Union, was really a political contradiction, for these two qualifications were simply incompatible. Stalin, in his bluff fashion, was more candid and consistent when he stated, "A freely elected government in any of these countries would be anti-Soviet, and that we cannot allow."[27] And the American political contradiction also entailed a grave moral responsibility, for it falsely encouraged some anti-Communist leaders to believe that their continued resistance would have solid American support. Had the United States government honestly conceded that it had no intention of incurring risks and sacrifices to challenge Soviet hegemony over Romania, some of these anti-Communists might have been able to save themselves, at least physically if not politically, while there was still time. Much has been written about alleged British perfidy toward the "London" Poles during and immediately after the war; but at least Churchill repeatedly (albeit vainly) urged his Polish allies to come to terms with Stalin before it was too late. This may not have been a particularly glorious page in British diplomatic history, but it was morally less culpable than the American posture in Romania and, as we shall see in the following section, Bulgaria of misleadingly exposing its protégés to very severe reprisals by fostering false hopes.

7

Whereas the Czechoslovak, Yugoslav, and Albanian Communist movements emerged from World War II far stronger than they had been during the interwar era, and the Polish, Hungarian, and Romanian Communist parties emerged intrinsically still as weak as they had been and utterly the creatures of their Soviet patrons, the Bulgarian Communists—uniquely in East Central Europe—emerged debilitated, a mere shadow of the powerful force that they had been in the interwar decades.[28] Some reasons for this enfeeblement were noted in Chapter 2. And their first political assignment from their Soviet protectors after the Fatherland Front *Putsch* of September 9, 1944—to popularize Bulgaria's belated military participation in the Soviet drive to clear the *Wehrmacht* out of southeastern Europe—only compounded the Communists' difficulties. The Bulgarian public, recalling that in April 1941 the Macedonian and Thracian irredentas had been redeemed cost-free under Axis sponsorship, now found it difficult to appreciate why it should sacrifice 32,000 dead soldiers and pay enormous costs to maintain a "fraternal" Soviet occupation army for the privilege of returning these acquisitions (which it perceived as rightful) to Yugoslavia and Greece—and without even gaining Allied cobelligerent status at the peace table for its pains. Soviet permission to retain Southern Dobruja, annexed from Romania in 1940, seemed poor compensation for these sacrifices of 1944. At least Romania's belated association with the Soviet armies had regained it northern Transylvania. Bulgaria, however, perceived itself to be a deep loser, and all the Communists' efforts to gild this bitter lily were in vain.

But though weak relative to their own erstwhile historical strength, the postwar Bulgarian Communists were stronger than any of their competitors, all of which were internally divided and none of which enjoyed equivalently sustained Great Power patronage. These competitors consisted of the minority sectors of the Agrarian and Social Democratic parties and the Zveno group of intellectuals and reserve officers—the Communists' three partners in the Fatherland Front—as well as two urban bourgeois parties: the Democratic, which had opposed the country's wartime affiliation with the Axis but had remained aloof from the Fatherland Front, and the Radical, which, though not in the wartime governments, had believed in German victory and thus could be all the more easily blackmailed by the Communists into deserting the opposition at a suitable moment.[29] The majority wings of the Agrarian and Social Democratic parties, which had declined to enter the wartime Fatherland Front, were now organizationally banned, as were all other parties of the old center and right.

It may plausibly be argued that it was precisely due to these real and perceived weaknesses of all the domestic players in the postwar Bulgarian political game that it was played with such extraordinary ferocity—even by conventional Balkan standards. The Communists' purge of their competitors started very early, was particularly violent, and was especially comprehensive, while the political courage of their outstanding opponents was also remarkably audacious and persistent. This struggle proceeded through five chronologically overlapping, yet analytically discrete, phases: (1) the destruction of the institutional and political pillars of the royal regime, (2) the emasculation of the non-Communist partners within the Fatherland Front, (3) the liquidation of the hitherto formally tolerated opposition to the Communists, (4) the internal purge of the Communist leadership cadres, and (5) full Stalinization.

Immediately following September 9, 1944, Fatherland Front committees, working with a new militia and with the political-police apparatus of the Interior Ministry (both dominated by Communists), proceeded to purge and monitor the formal governmental structure. This process was accompanied by mass trials and executions of several thousand real and alleged fascists-cum-war criminals, in the course of which many personal and political scores were arbitrarily settled as the victims ranged from the wartime regents, ministers, and parliamentary deputies down to village mayors, policemen, and tax collectors. Rumor estimated the number summarily executed at between 20,000 and 100,000.[30] Even allowing for exaggeration at the upper margin of these rumors, the entire process was so brutal and exaggerated as to provoke a deep anti-Communist revulsion, leading to reconciliations in the divided Agrarian and Social Democratic movements and hence to the emergence of a more or less coordinated anti-Communist opposition within and without the Fatherland Front.

In a maneuver resembling (albeit preceding) the Hungarian Communists' intervention to ensure that the Smallholder party's leader be the more pliable Ferenc Nagy rather than the stouter Dezső Sulyok (see section 4), the Soviet deputy chairman of the Allied Control Commission for Bulgaria, General Sergei S. Biryuzov, forced Georgi M. Dimitrov to relinquish the leadership of the Bulgarian Agrarian Union to Nikola Petkov in January 1945. Dimitrov (conventionally nicknamed "Gemeto" after his initials and to distinguish him from a Communist leader with the same given name) had spent the war years under British protection in the Middle East and hence was suspect as too pro-Western. But in Bulgaria, in contrast to the Hungarian scenario, the maneuver backfired because Petkov, contrary to expectations, grew into a tenacious adversary of the Communists' drive to power, first inside the Fatherland Front and then, following his expulsion from it after a spurious, Communist-arranged palace coup in his own Agrarian Union in May 1945, outside it. A similar manipulated self-mutilation was inflicted on the Social Democrats shortly thereafter. Thus already by the autumn of 1945, the originally authentic wartime Fatherland Front coalition had been rendered bogus.

Precisely because he was a genuine radical and a true believer in Bulgarian–Russian friendship, Petkov's resistance to the Communists carried great moral authority. And because both his father and his older brother had been the victims of political assassinations, he appeared to be imbued with an almost fatalistic fearlessness. As his popularity and the readership of his independent newspaper soared and Communist morale correspondingly faltered in the autumn of 1945, Moscow felt constrained to intercede. First, the internationally renowned veteran Bulgarian Communists Georgi Dimitrov and Vasil Kolarov, who for over two decades had made their careers in the Soviet Union and the Comintern apparatus, were dispatched back home to dim Petkov's luster. Then, Soviet Deputy Foreign Minister Vyshinsky spent January 9 to 12, 1946, in Sofia, pleading with Petkov and his independent Social Democratic colleague Kosta Lulchev to reenter the Fatherland Front government and thereby reburnish its tarnished legitimacy. It will be recalled that a similar broadening of the Groza government in Romania was arranged at this time, following the Moscow Conference of the Big Three foreign ministers, but in time proved merely cosmetic (see section 6). Petkov and Lulchev were made of sterner stuff than their Romanian counterparts, countering Vyshinsky's petition with demands for (1) the retroactive cancellation of parliamentary elections held the previous November 18, which they had declared to be rigged and had boycotted, and which the Fatherland Front declared itself to have won with 88 percent of the ballots, and (2) the surrender by the Communists of the ministries of Interior and Justice, the key bureaucratic instruments of control and purge. The Bulgarian Communists were by now so shaken by the public backlash to their manipulations and excesses that they were prepared to meet the opposition halfway, reluctantly agreeing to yield to it the Justice portfolio and to let it designate two undersecretaries in the Interior Ministry. But Moscow, perhaps concerned that such concessions would prove contagious to Communist morale elsewhere in East Central Europe, vetoed this tentative compromise in March 1946.

As Petkov and Lulchev were emphatically supported and encouraged by the American political representative in Bulgaria, Maynard B. Barnes, their harassment eased temporarily, and the Communists concentrated during the spring and summer of 1946 on purging their Zveno allies and the army officer corps. The Zveno war minister was dismissed on the pretext of his prewar professional contacts with the Yugoslav Draža Mihajlović, who was being prepared for trial by the Tito regime in Belgrade, and his functions were transferred to the cabinet as a whole. But he was gently reassigned to be envoy to Switzerland. The Zveno prime minister was downgraded to vice premier and foreign minister, yet also physically spared. As long as the peace treaty was not signed and Western diplomatic recognition had not been granted, prudence suggested that the British and Americans not be gratuitously irritated by a resumption of brutality. And Zveno reciprocated by never defying the Communists, as had the Agrarians and the Social Democrats.

Although terrorist excesses might be suspended, the Communists' political drive to power continued apace. Two thousand officers of the royal army were retired in the summer of 1946. On September 8 (the anniversary eve of the Fatherland Front *Putsch* of 1944), a plebiscite replaced the monarchy with a republic, the adolescent Tsar Simeon II being exiled and compensated with a fairly generous gratuity of $20 million. The official results of this plebiscite were 92.3 percent for a republic, 4.8 percent to retain the monarchy, and 2.9 percent invalid ballots. On October 27, 1946, a constituent assembly was elected to draft a people's democratic republican constitution, which was adopted on December 4, 1947. Of the 465 seats at stake, the Fatherland Front won 366, of which 275 were assigned to the Communist party, and 91 to its subordinate partners. During the next year, however, 27 of the non-Communist Fatherland Front deputies defected to the opposition, which had originally emerged from the elections composed of 89 Petkovist Agrarians, 9 Lulchevist Social Democrats, and 1 independent. In percentage terms, the Fatherland Front had claimed 70 percent of the ballots for itself and conceded 30 percent to the opposition; this was the largest proportion recorded for any real opposition in any postwar East Central European election, and tended to support Petkov's claim that had the balloting been truly free, secret, and accurately counted, he would have won with 60 percent popular support.

Now ensued a shameful Western failure of nerve. Even though as late as November 4, 1946, Washington and London had endorsed Petkov's criticism of the elections' procedural pollution through intimidation, even though he and Lulchev remained undaunted, and even though fissures were now showing in the Fatherland Front as some of the Communists' hitherto tame allies were straddling and deserting, the United States disarmed itself of its only leverage for influencing Bulgarian domestic developments by signing the peace treaty on February 10, the Senate then ratifying it on June 4, 1947. That is all the Communists and the Soviets needed. The day after this American ratification, Petkov was arrested on the floor of the parliament. His newspaper had been shut since April through the sinister device of an ostensible printers' strike. Tried during August in an atmosphere of drummed-up hysteria on charges of criminal conspiracy to overthrow the government, he was found guilty and hanged on September 23, his body relegated to an unmarked grave. Prime Minister and Communist party leader Georgi Dimitrov, who had been the beneficiary of a world-wide protest movement at the time of his trial in Nazi Germany in 1933 for having been allegedly involved in the burning of the Reichstag, now had the odious taste to announce that, but for "provocative" British and American protest notes, Petkov's death sentence would have been commuted. Hardly was Petkov's body cold in the ground than the Americans extended formal diplomatic recognition and designated an ambassador to Bulgaria on October 1, 1947.

What was disgraceful in this episode was not Washington's decision to draw the "containment" line in this part of the world around Greece and Turkey and to write off Bulgaria. That decision can be defended as, on balance, prudent and rational. But not the failure to signal it in time to Barnes in Sofia and to restrain him from fostering Petkov's belief that his American backing was solid—a belief that lured Petkov into actions forfeiting his life.[31]

The rest was anticlimactic, though ruthless. Petkov's Independent

Agrarian Union had been dissolved and its parliamentary mandates canceled immediately after his trial. With impressive pluck, the nine opposition Social Democratic deputies still voted against the Fatherland Front's constitutional draft in December 1947 and against its proposed budget in January 1948. They were arrested in July 1948 (one managing to flee to Turkey) and sentenced to long prison terms in November. In August, the puppet Social Democrats within the Fatherland Front dissolved their nominally autonomous party and merged with the Communist party. The previous month, the puppet Agrarian Union within the Fatherland Front had announced its mutation from a "political" into an "educational" body. In early 1949, Zveno and the Radical party ended their own organizational existences altogether and dissolved themselves into the Fatherland Front as a whole, which soon became simply a mass association encompassing virtually the entire population.

At the end of 1948, over 60 percent of the national income was still in private hands, even after some fairly extensive nationalizations during the two preceding years. This changed with the launching and implementation of two successive five-year plans starting on January 1, 1949. By their close, private ownership in industry and trade had vanished, and agriculture had been completely collectivized. Despite the imposition of ferociously exacting labor discipline, productivity in many economic sectors remained below prewar levels. Yet even during and since the post-Stalin "thaw," the Bulgarian regime made fewer structural socioeconomic concessions to its peasantry and petite bourgeoisie than did any other people's democracy.

The death of Georgi Dimitrov on July 2, 1949, in Moscow occurred in the midst of a bloody internal purge that devastated the Communist elite and cadres over the next year and that will be analyzed in Chapter 4. Though by this time, Dimitrov's standing in Stalin's eyes was none too secure, his opportune demise facilitated the inauguration of a cult-legend to his memory that functioned as both a tool in and a fig leaf over this sanguinary purge.

8

In Albania, as in Yugoslavia, the Communist capture of political power had been completed during the war (see Chapter 2, section 8). The collateral structural transformation of this relatively backward society was then pursued in January 1945 with the imposition of a confiscatory "war profits tax" that crippled the small native bourgeoisie and the nationalization of all Italian and German assets, whereby the country's industrial and transportation stocks came into state ownership. The peasantry—the most numerous class—was conciliated, while the powerful landowners were broken with a series of decrees canceling agricultural debts, slashing land rents by three-quarters, and nationalizing water resources. Yet though determined to consolidate their exclusive hold on power, the Albanian Communists at this early stage still craved international, including British and American, recognition and respectability. Among the cosmetic gestures that they arranged to elicit this legitimation were elections to a constitutent assembly, held on December 2, 1945, in which 89.8 percent of the eligible voters participated, of whom 93.2 percent endorsed the uncontested list of the Communist-controlled Democratic Front. As in the previous month's elections in Yugoslavia, particularly bold dissenters had the dubious option of the "opposition urn." On January 11, 1946, this assembly abolished the monarchy of King Zog and proclaimed the people's republic; on March 14, it adopted a constitution copied from the contemporaneous Yugoslav one, minus its federalism.

The triadic interaction within the Albanian Communist movement of (1) the equivocal stabs toward earning recognition and possible aid from Great Britain and the United States, (2) the emphatic linkage, forged during the war, to Tito's Yugoslavia, and (3) domestic politics, was quite complicated. The West, discounting the formal Albanian gestures as transparently contradicted by the hard political reality of the Communists' heavy hand, sent signals that appeared to endorse ongoing Greek aspirations to annex southern Albania (northern Epirus). The Communists, predictably, now exploited this incipient threat to present themselves as passionate protectors of Albanian national integrity, a pose that was, however, marred by their passive return to the Titoists of the Albanian-populated region of Yugoslavia that the Axis had assigned to Albania in 1941. Compounding this ambiguity was the fact that the Communists were already much weaker among the Geg clans of north-central Albania (Abaz Kupi's base during the war), who had fraternal ties to the Albanians in Yugoslavia, than among the southern Tosk Albanians, who were more threatened by Greek revisionism. In effect, therefore, the Albanian Communists' tight nexus to the Yugoslav Titoists, who were just then in a virtual paroxysm of hypermilitancy and ultraradicalism, was proving to be a double embarrassment, aggravating their international isolation and alienating the northern population.

While the latter embarrassment was, in a sense, "resolved" by the tried and tested techniques of terror and propaganda by which the Communists imposed their authority on the north, the former one tore the Communist party apart. One must appreciate that the Yugoslav embrace was so smothering that it isolated Albania even from other Communist-dominated countries, including the Soviet Union (which in May 1946 declined to receive a top-level Albanian governmental delegation), as well as from the West. Only with Yugoslavia did Albania have treaty relations (as of July 9, 1946), and the Yugoslav Communist party even represented the Albanian one at the founding meeting of the Cominform, held from September 22 to 27, 1947. Meanwhile, Yugoslav advisers and specialists were flooding into Albania's administrative and military cadres, while Yugoslav investments, loans, and monetary and customs arrangements virtually annexed the economy. The relationship became ironically reminiscent of the prewar one with Mussolini's Italy. Nevertheless, one wing of the Albanian leadership, led by Koçi Xoxe and composed mainly of men of working-class background, had no reservations about this dependence and was even prepared to take Albania into Yugoslavia as a seventh republic. The other wing, consisting primarily of intelligentsia and led by Enver Hoxha and Sejfulla Malëshova, wanted to pursue a more independent and broader course, both internationally and domestically. In the winter of 1945/46, Xoxe liquidated Malëshova, with Hoxha acquiescing but uneasy. Hoxha's opportunity to turn the tables came within the broader setting of Stalin's repudiation of Tito. An early hint of this was a state visit in December 1947 by Hoxha to Sofia, where, with Soviet approval but to Yugoslav chagrin, he broke Albania's diplomatic isolation by signing a treaty with Bulgaria. Then, three days after the publication of Stalin's (formally the Cominform's) repudiation of Tito on June 28, 1948, Hoxha denounced all economic arrangements with Yugoslavia as being incompatible with Albania's sovereignty and ordered the Yugoslav personnel to leave within forty-eight hours. The Soviet Union and several of its satellites promptly compensated Albania for the forfeited Yugoslav aid with generous subsidies. Xoxe and his pro-Tito associates were stripped of their governmental offices in October 1948, expelled from the party in November, and executed for treason on June 11, 1949. (Unlike other purge victims in other people's democracies, he was never posthumously exonerated or rehabilitated.)

Stalin's support thus enabled Hoxha not only to eliminate his personal domestic rivals, but also to gain for Albania a more powerful and more remote patron in exchange for a proximate and directly threatening one. Stalin, reciprocally, gained a direct naval base in the Mediterranean and the utter isolation of Tito within the world Communist movement. Interestingly, the lever that pried open both the Yugoslav–Soviet and the Albanian–Yugoslav rifts was an amalgam of nationalism and the craving of individuals for personal political control.

Hoxha quickly became an adept student of Stalinist techniques of rule. His regime became so oppressive and vindictive—albeit stoutly committed to economic development and modernization—that Western intelligence services assessed that it might be overthrown by subversion. But all their schemes (whose chances of success were, in any event, quite dubious) were betrayed by the Soviet mole in the British Secret Intelligence Service, H. A. R. (Kim) Philby.[32]

9

Two generalizations—one historical and moral, the other political and analytic—emerge as appropriate conclusions from and to this chapter. The first is that the Communist capture of power, for all its imposition by the Soviets, was greatly facilitated by the fact that, domestically, the Great Depression and World War II not only had destroyed the old political systems, but also had gravely weakened the old political classes throughout East Central Europe. The Communist cadres that replaced those traditional political classes initially capitalized on a widespread craving for change, but quickly disillusioned, soured, and indeed reversed that very craving. They offended the societies over which they ruled not so much by monopolizing power—after all, rule by "government parties" was quite conventional in this part of the world—but by abusing it beyond traditional or acceptable limits and by putting it at the service of another state and society, the Soviet Union, in which they appeared to place their ultimate loyalty. This impression seemed to be corroborated by the purges of the late 1940s and early 1950s, which will be treated in Chapter 4.

The second generalization is that, except in Yugoslavia and Albania, the first postwar governments in the states of East Central Europe were coalitions, with the Communists formally but one of several partners. Since it is difficult for trained Communists to be members of a coalition without seeking to dominate it, they would defend these coalitions against external critics while subverting them from within. This subversion was facilitated by the fact that all these coalitions were larger than they needed to be for reliable governmental majorities. And it is a truism of political science that an oversize multiparty coalition, in which some members are arithmetically and politically superfluous, will be subject to strains as each member maneuvers to avoid being targeted as redundant and to remain among the surviving partners.[33] Hypothetically, the Communists might have been deemed expendable by smaller, tighter coalitions of peasant, Socialist, and/or liberal parties. (In Hungary, indeed, the Smallholders even emerged with an absolute majority from the elections of November 4, 1945.) But this contingency was a priori excluded by Soviet fiat. Given, then, the imposed indispensability of the Communists, each of the other coalition partners tended to insure itself against becoming the extruded party by muting its grievances and postponing its demands, until either the moment was too late or the issue chosen to make a stand was inappropriate. In the process, many of these parties split into collaborating and opposition fragments under the stress of these pressures and maneuvers—all grist for the Communist mills. Finally, since the concepts of a loyal opposition and an open society are alien to the Leninist-Stalinist mind-set, it followed that, once they had broken their erstwhile coalition partners, the Communists would try to preclude any possibility of a freely competitive political system reversing their "historic victory." They would seek to do this not only by monopolizing political power, but also by transforming the socioeconomic "substructure" in an ideologically stipulated direction.[34]

AUSTRIA
HUNGARY
SLOVENIA
IT.
Ljubljana
Trieste
Zagreb
VOJVODINA
ROMANIA
CROATIA
Novi Sad
Belgrade
BOSNIA-HERCEGOVINA
ADRIATIC
Sarajevo
SERBIA
YUGOSLAVIA
International boundary
Republic boundary
Autonomous province boundary
National capital
Republic or autonomous province capital
City
0 25 50 75 100 kilometers
0 25 50 75 100 miles
MONTENEGRO
Priština
BULGARIA
Titograd
KOSOVO
SEA
Prizren
Skopje
ALBANIA
MACEDONIA
Tirana
GREECE

4

The Dialectics of Stalinism and Titoism

1

Within two or three years of the end of World War II, each Communist party in East Central Europe was well on the way toward capturing and/or consolidating political power in its country. Nevertheless, from Moscow's perspective, the overall picture was still one of excessive diversity. Not only did the pace of establishing Communist rule, and its comprehensiveness, differ from state to state, but several of the Communist-dominated regimes were at loggerheads as a result of having succumbed to "national domestic" perspectives. Thus the Polish and East German Communists were making discrepant statements about the permanence of the Oder-Neisse frontier, the Polish and Czechoslovak ones about their states' respective border claims in the Teschen (Těšín, Cieszyn) and Glatz (Kladsko, Kłodzko) districts, the Czechoslovak and Hungarian ones about the treatment of the Magyar ethnic minority in Slovakia, the Hungarian and Romanian ones about the rectification of the frontier in Transylvania, and the Yugoslav and Bulgarian (as well as the Greek) ones about the just disposition of Macedonia. "Domesticism" had also characterized the diverse initial responses of the East Central European governments to the American Marshall Plan overture in June 1947. Most egregiously, Tito was pressing Yugoslav national claims to Trieste so belligerently and with such a cocksure assumption that the Soviet Union had no choice but to back him to the hilt—even to the brink of general war—against British and American support for Italy's counterclaims that Moscow showed concern.[1] Diversity, indiscipline, domesticism, and the craving of some people's democratic tails to wag the Soviet dog appeared to be running rampant. For a political system such as Leninism-Stalinism, with its ideologically ingrained distaste for spontaneity, effervescence, pluralism, and factionalism, such a condition was likely to be deemed intolerable ere long, even in the best of circumstances. All the more so now, in 1947 and 1948, as the Cold War was descending on Europe in earnest. Stalin's method for imposing political and structural order on the East Central European Communists was characteristic and reminiscent of the tactics by which he had consolidated his personal mastery of the Soviet Union in the decade after Lenin's death. First, he created the Cominform in September 1947, using the Yugoslavs as his hatchet men to discipline and bully the more laggard Communist parties. Then he turned around and used this same instrument, the Cominform, in his effort to liquidate the excessively independent and rambunctious Titoists.

Delegates from nine Communist parties—the Soviet, Polish, Czechoslovak, Hungarian, Yugoslav, Romanian, Bulgarian, French, and Italian—attended the founding session of the Cominform (Communist Information Bureau) in the Polish Silesian spa of Szklarska Poręba, on September 22 to 27, 1947. The hitherto acceptable notion of distinctive national paths to socialism (which had left the pace and forms of political and socioeconomic transformation to local discretion) was rejected, the Czechoslovak comrades berated for their failure to seize exclusive power (which they would correct the following February), and the French and Italian parties humiliated for having permitted themselves to be extruded from their erstwhile coalition governments without recourse to civil war. In this orgy of recrimination and whip cracking, the Yugoslav delegates led the pack in collusion with the Soviet ones.[2] Resolving that the major current error of the Communist parties would be to underestimate their own power and overrate the strength of their foes, the first Cominform conference was quickly followed by an acceleration in the pace of the Communist monopolization of power throughout East Central Europe and by a corresponding wave of violent, Communist-instigated strikes in Western Europe. In the people's democracies, the bourgeois and agrarian parties were now broken, the Social Democratic ones forcibly absorbed, and their respective leaders arrested, executed, or forced to flee abroad. Constitutions imitating the contemporaneous Soviet one were adopted between 1948 and 1952 in all the people's democracies. Indeed, the very concept of "people's democracy," which had implied a certain respect for diverse national particularities, was radically revised to emphasize its uniform organic dependence on the Soviet exemplar.

In terms of policy and administration, the new emphasis on discipline, uniformity, Communist monopoly, and Soviet priority entailed drives to collectivize agriculture and to press rapid industrialization, with a distortive overemphasis on heavy metallurgical industries. The several economies of the East Central European people's democracies were linked to that of the Soviet Union through a series of bilateral arrangements, but were deliberately discouraged from regional integration, just as Stalin prohibited multilateral political discussions once the Cominform had been formally launched. Fearing the cumulative weight of a regional combination, he preferred each people's democracy to be exclusively dependent on the Soviet Union. Though economic development in East Central Europe was undoubtedly achieved, albeit in an autarkic, unintegrated, and ultimately irrational manner, hardship, terror, bureaucratic hypertrophy, elite arbitrariness, political servility, and cultural aridity were the concomitants of these Stalinist policies and outlooks.

Of course, the more the Soviet Union insisted on uniformity, radicalization, imitation of itself, and unquestioning acknowledgment of its imperial priorities in East Central Europe, the more it was obliged to intervene explicitly in the domestic affairs of the area's states and to seek direct administrative control of their Communist parties. This proclivity, however, brought it into conflict with Tito's Yugoslav Communists, who, their own radicalism and admiration of the Soviet model notwithstanding, treasured even more their autonomy and cohesion, hard-won as they had been during the bitter wartime Partisan struggles.

2

A certain degree of implicit, unacknowledged tension had characterized the relationship between Tito and Stalin during World War II, when the Soviets had been far freer with political advice than with material help to the Partisans, and when the thrust of that (ignored) advice had been to try to dissuade them from exploiting the war situation to seize power from the royalist–Četnik forces (see Chapter 2, section 5). With the war over, however, that tension (which had been screened from the intermediate cadres and the general rank and file) had apparently been transcended, as Tito moved rapidly to distance himself from his wartime British and American suppliers of arms and to establish in Yugoslavia the earliest and most closely imitative Stalin-type political system (see Chapter 3, section 5). In the long run, though, this vivid reconciliation proved fragile, for Stalin was less impressed by the scenario of close and voluntary imitation than he was concerned with the autonomy of the imitator. Aggravating this basic flaw in the relationship were Tito's provocative, premature, and risky (in Stalin's judgment) anti-Western Cold War behavior in 1945 and 1946 and his posturing on grandiose state visits to other East Central European capitals during those years, as though he were Stalin's designated viceroy for the area, especially its southern Danubian and Balkan half. And whenever they met in Moscow, Tito—while treating Stalin respectfully—would convey that he had come to negotiate, not to receive orders, and that he was a partner, not an instrument. The specific accusations that were exchanged after their rift became public in 1948 read like a glossary to this fundamental issue of the Titoists' structural independence and their concomitant independent behavior.

During and immediately after the war, the membership of the Yugoslav Communist party had increased exponentially from 12,000 in 1941 (of whom 3,000 survived the war) to 140,000 in 1945 and 470,000 in 1948. The new members had known no leader but Tito. The party was thus politically and administratively the creation of Tito and his inner circle. Furthermore, thanks to the nature of the wartime Partisan movement, this party had total control over the country's military and security forces, and hence was able to thwart postwar Soviet efforts to infiltrate and subvert them. Indeed, the behavior of the Soviet military advisers and technicians who were seconded to the Yugoslav armed forces between 1945 and 1948 was by itself counterproductive to such Soviet efforts, for they were overpaid, arrogant, and woundingly contemptuous of the Partisans' achievements and traditions. They thus compounded the unfortunate impression that had been made by the rapine and drunkenness of the Soviet army during its combat passage through Yugoslavia in 1944 and 1945. Yet another irritant in the military dimension of the relationship was Stalin's indirect attempt to curb Tito's foreign-policy adventurism by restricting the supply of spare parts for the armaments that the Soviet Union was supplying to the Yugoslav armed forces—armaments that the Yugoslavs deemed obsolescent.

In the economic dimension, the Yugoslav planners, heedless of the Soviet Union's own wartime losses, blandly assumed that it was bound to subsidize their overambitious industrialization program, which called for the doubling of the national income by 1951 over its 1939 level, with an increase of 223 percent in industrial and of 52 percent in agricultural outputs, all based on a reinvestment rate of between 25 and 33 percent of the national income. When the appalled Soviets recommended moderating the planned, and indeed reckless, pace of industrialization while concentrating instead on increasing agricultural and raw-materials (mining, lumbering) production, the offended Yugoslavs (with but two leaders dissenting) retorted that such a policy would relegate them to continued semicolonial dependence on the advanced economies of the world. And, indeed, the price pattern of Soviet–Yugoslav trade did generally disadvantage the smaller and weaker partner. More specifically, the Yugoslavs halted the proliferation of joint companies with the Soviets after February 1947, when two were established and were quickly deemed to be exploitative and discriminatory.

In the area of foreign policy as well, Tito's adventurism and hubris dismayed Stalin, while Stalin's assignment of priority to his relations with the other two Great Powers annoyed Tito. In addition to the irritants already alluded to, there were Tito's manifest, albeit unacknowledged, intervention on the Communist side in the Greek Civil War (which Stalin viewed as an unpromising and gratuitous provocation of the British and Americans) and his patronage of projects for a large Balkan federation to include Hungary, Romania, Yugoslavia, Bulgaria, Albania, and an eventually Communist Greece (which apparently struck Stalin as a plot to build a power bloc that would challenge Soviet hegemony). The Macedonian question, which had bedeviled relations among Yugoslavia, Bulgaria, and Greece before and during World War II (see Chapter 2, sections 5 and 7; Chapter 3, section 7), continued to function as an exacerbating "hidden agenda" in the different Balkan-federation schemes. An irritant ancillary to foreign policy was the attempt by Soviet diplomatic personnel assigned to Yugoslavia to pry into the affairs of the Yugoslav party and by Soviet intelligence agents masquerading as diplomatic personnel to recruit Yugoslav citizens, and the reciprocal surveillance of these Soviet diplomats and pseudo-diplomats by the Yugoslav security services.

It is remarkable that though they were clear-eyed and even scrappy in noting each of these many irritants as they arose, the Yugoslavs utterly failed to anticipate their cumulative impact on Stalin's mind, never appreciating that for him the issue was not the rights or wrongs of this or that disagreement, but the compound evidence of the refractory autonomy of the Yugoslav actor. Hence as late as September 1947, they still myopically allowed themselves to be "set up" as the Soviet party's hatchet men vis-à-vis more moderate parties at the founding of the Cominform. And when Stalin subsequently decided to break with them, the Yugoslavs could not, for a long time, bring themselves to accept the finality of his decision, insisting that the problem was the product of misreportings and misunderstandings susceptible to correction and resolution. Yet Tito, who had been in Moscow at the height of the Great Purges in 1937, was under no illusion that recanting—be it ever so abjectly—would gain him personal absolution. It is important, in this connection, to bear in mind that Stalin broke with Tito, not vice versa.

At a trilateral meeting in the Kremlin of Soviet, Yugoslav, and Bulgarian party delegations on February 10, 1948, Stalin was rude and unfriendly, complained of being kept in the dark about the Yugoslavs' wider Balkan-federation designs, and ordered an immediate small federation of Yugoslavia and Bulgaria. As the Bulgarian Communists were notoriously amenable to Soviet control, Tito viewed this pressure as an attempt to insert a Trojan horse into the cohesive, Partisan-steeled Yugoslav cadres. He alerted his own Central Committee to this apprehension on March 1, simultaneously signaling his concern about Soviet apathy toward Yugoslavia's industrialization plan and apparent intent to reduce its economy to extractive dependence.

On March 18 and 19, 1948, the Soviet Union peremptorily recalled its military and civilian advisers from Yugoslavia, alleging, in a shrill letter of March 27, that these experts had been treated with hostility and abuse, the Yugoslav Communist party had succumed to multiple ideological heresies all of which were spelled out, it was riddled with British spies, its cadres were intimidated by the state security services (secret police), and its leadership was illegitimately self–co-opted rather than democratically elected. (As though the last two allegations were not a fortiori applicable to the Soviet Communist party itself.) The long and detailed Yugoslav reply of April 13 made four basic points: (1) Stalin is the victim of false information sent by his meddling and snooping diplomats and other agents in Yugoslavia who, in turn, draw these mendacities from a few irresponsible "antiparty" elements in the Yugoslav Central Committee and from illicitly suborned functionaries; (2) not only is our party not intimidated by the security services, but also is internally cohesive and comradely and far more solidly and creatively in control of the country than are the Communist parties of neighboring states; (3) our party has inculcated love of the Soviet Union among the masses, which has been neither easy nor automatic, and though we do love the Soviet Union, we are entitled to love our own country no less; (4) we are building socialism differently from, but no less legitimately than, the Soviet Union, and our model, too, has international revolutionary validity for other countries. The unrepentent tone and (in the context of the prevailing dual cults of Stalin and of the Soviet Union in world Communism) the politico-ideological hubris of these retorts and claims provoked a somewhat frenzied, repetitive Soviet screed of May 4, stigmatizing the Yugoslav stance as "childish," "laughable," and "arrogant"; insisting that the behavior of Soviet personnel in and Soviet policy toward the people's democracies was not bound by or comparable with the formal rules and modes that characterize relations among bourgeois states; and claiming that the Yugoslav Communists were in power in their country only thanks to the military prowess of the Soviet army and not to their own Partisan exertions.

By thus belittling the wartime Partisan sacrifices, Stalin committed a grievous blunder, on which Tito promptly capitalized to rally his party and his people to his side. A corollary Soviet gaffe was to publish the accusations and to announce the expulsion of the Yugoslav Communist party from the Cominform on June 28 (Vidovdan, or St. Vitus's Day), a day on which Yugoslavs are likely to feel particularly patriotic and exalted because it is the multiple anniversary of the epic Serbian national catharsis at the Battle of Kosovo against the Ottomans in 1389, of the assassination of the Habsburg Archduke Franz Ferdinand at Sarajevo in 1914 (an event regarded by Yugoslavs as a valiant and liberating deed), and of the adoption of modern Yugoslavia's first constitution in 1921. But Stalin had landed one telling, albeit sly, rhetorical blow by indicating, on May 22, that the Yugoslavs' refusal to attend a second Cominform plenum, at which their own deviations were on the agenda, contrasted shabbily with their alacrity in leading the pack to criticize the moderate Communist parties at Szklarska Poręba the previous September. The Cominform's expulsion communiqué restated all the Soviet claims and charges from the preceding correspondence (which had been signed by Stalin and Molotov on the one side and by Tito and Edvard Kardelj on the other) and closed with an appeal to the "healthy elements" in the Yugoslav Communist party to replace its current leaders.

Why this replacement of leaders did not happen is clear: Tito's party, having conquered power by its own efforts, was cohesive, confident, comradely yet centralized, tightly structured, recruited and shaped by himself, totally in control of its country and its state apparatus, broadly polyethnic, and free of that fissure between "local undergrounders" and "Muscovites" that plagued those Communist parties that had been installed in power by Soviet leverage; Tito was indeed a good Stalinist in his appreciation that in politics, organization is as important as policy. *That* it did not happen surprised and bewildered Stalin, exposing his grave miscalculation. He appears to have believed that "all I need to do is shake my little finger and there will be no more Tito."[3] Initially, he expected that an internal leadership *Putsch* would oust Tito. But the pair on whom he counted for this, Andrija Hebrang (a Croat) and Sreten Žujović (a Serb), were easily isolated, expelled, and arrested in May and June. He then seems to have toyed with the notion of a military coup or cross-border harassment by exiles who were loyal to Moscow. But this, too, was aborted when three senior Yugoslav officers on whom Stalin had pinned this hope were intercepted while trying to cross into Romania in mid-August. A propaganda barrage proved similarly futile and, indeed, counterproductive because Tito thereupon published the previous spring's correspondence, replete with Stalin's hurtful belittlement of the Partisans' wartime heroism. Economic warfare was equally unavailing, as the Yugoslavs maneuvered dexterously and the West behaved wisely in blunting it (see Chapter 3, section 5). The really hard question is why Stalin did not order an outright invasion of Yugoslavia, albeit against fierce resistance, by the Soviet and other people's democratic armies between 1948 and 1950, before the Americans rearmed in response to the outbreak of the Korean War. After all, Stalin's need to destroy Tito was now very high, since the aura of his own authority was so heavily a function of his undefiability.

3

Precisely in order to prevent the wound inflicted on his charisma by Tito's defiance from hemorrhaging into a generalized series of defections and challenges, Stalin now recoiled into an orgy of purges in the other people's democracies and of political constrictions in the Soviet Union itself. Such purges might well have been initiated sooner or later in any event, but their timing, character, and thrust were strongly affected by Tito's defiance, in whose wake the need to demonstrate Soviet power and Stalinist infallibility became all the more imperative. Contrary to what has often been suggested, the scenarios of these purges in the people's democracies are not all subsumable under a single formula, such as Stalin's arranging for the wartime "Muscovites" to eliminate the "local undergrounders" or authorizing the Communists who belonged genealogically to the eponymous state-nations to liquidate the Jewish and other ethnic-minority ones. The actual tapestry of the East Central European purges, taken as a whole, was more variegated than such simplifications would allow, though the net outcome everywhere was to render the surviving Communist leaders of the "loyal" people's democracies dependent, insecure, and utterly subservient to Moscow for the rest of Stalin's lifetime. As the Czechoslovak daily *Rude Pravo* (*Red Truth*) was to warn on May 25, 1952, "Love of the Soviet Union does not tolerate the slightest reservation. The road into the morass of treason begins on the inclined plane of reservations and doubts regarding the correctness of the policy of the Soviet Union."[4]

It may be true, as the Yugoslavs later claimed, that before the schism erupted publicly, several of the Communist leaders in neighboring states had clandestinely sympathized with Tito, who, after all, was one of them and, like some of them, sought to get the Soviets off his back.[5] But it is also likely that many of them resented his condescending airs and his preceptorial manner and longed to see him taken down a peg. Once the Soviets had signaled their refusal to compromise, sheer self-interest probably prompted all the neighboring leaders to hope for Tito's early collapse or capitulation, since any prolongation of his resistance was bound to draw Stalin's baleful eye to their satrapies severally and collectively, with all their messy diversity, indiscipline, and domesticism during the years 1944 to 1948.[6]

For strategic, demographic, historical, and economic reasons, Stalin appears to have viewed Poland as his most important and most difficult client state. And the manner in which he imposed Communist rule on Poland between 1944 and 1948 probably was the single most consequential precipitant to the breakup of his wartime alliance with the British and Americans and its succession by the Cold War (see Chapter 3, section 2). Because he did not permit even a brief postwar democratic interlude in Poland, such as was allowed in Czechoslovakia and Hungary, an erroneous impression has gained currency that Communist and Soviet behavior was more brutal in Poland than elsewhere in East Central Europe. But this is not quite true, at any rate not across all nuances of the political spectrum. As has been shown, Stefan Cardinal Wyszyński was treated more gently than were clergymen in several other people's democracies; traditional Polish military uniforms and other national symbols were uniquely retained even under Konstantin K.

Rokossovsky's tenure as defense minister; the peasantry was approached more gingerly than elsewhere; and, as we shall now see, the so-called Polish Titoists were dealt with more civilly and less violently than were the victims in any neighboring purge.

Władysław Gomułka had become leader of the underground Communist resistance in 1943 after the violent elimination of two predecessors under murky circumstances. He was confirmed as secretary-general of the party after the liberation, when he also became a deputy premier and the minister for the regained (ex-German) territories, in which capacity he administered the party's extensive patronage apparatus for the distribution of these new lands to resettled Poles. Together with a small cluster of fellow "undergrounders," he soon found himself at odds with the party's "Muscovites" over a number of policy issues and general stances. Gomułka tended to emphasize a "Polish national road" to socialism, to limit imitation of and dependence on the Soviet Union, to respect the patriotic traditions of the Polish Socialists, and to depreciate the collectivization of agriculture. This pattern of preferences and priorities could easily be labeled Titoist, and when he persisted in cleaving to it after mid-1948, even in the teeth of the new Stalinist and Cominform insistence on uniformity, emulation of Soviet patterns, socioeconomic radicalization, and unilateral political monopolization, Gomułka became a marked man. At a tense session of the party's Central Committee, held from August 31 to September 3, 1948, he was accused of multiple "rightist and nationalist deviations" and was removed as secretary-general. Ouster from his ministerial assignments followed on January 21, 1949, expulsion from the Central Committee on November 13, 1949, and arrest on August 2, 1951. But no total, abject confession to truly heinous crimes and sins, such as was wrung from the victims of intraparty purges in neighboring states, was extracted from Gomułka; nor were he and his fellow Polish "deviationists" (some of whom were less courageous and stubborn than he) subjected to the notorious show trials and executions that Stalin required of the other people's democracies. Indeed, Gomułka and his associates survived to be freed in 1954 and 1955, rehabilitated into the party in 1956, and dramatically reelevated to power during that year's October crisis.

Though it was widely, albeit surreptitiously, noted that Gomułka and his group were ethnic Poles, whereas their "Muscovite" purgers included a number of Polish Jews, overt anti-Semitism had not been projected by either side in this intraparty rift. Quite otherwise was the case in Czechoslovakia. A second significant difference is that whereas the Polish purge involved real disagreements over important policy issues, the Czechoslovak one was a raw power struggle among ambitious and insecure persons, exacerbated to a hysterical pitch by Stalinist scheming. And third, the outcome in Czechoslovakia (and elsewhere) was to be far more lethal than in Poland.

Rudolf Slánský had been the secretary-general of the Czechoslovak Communist party since 1945. A professional Communist functionary from his youth, a "Muscovite" during the war (when he was parachuted into insurgent Slovakia in September 1944), efficient, cold, narrow, suspicious, respected but not liked in the party, and Jewish, he was generally regarded as being politically devoted to the more gregarious, popular, earthy, proletarian, and ethnically Czech party chairman (and state president after June 1948) Klement Gottwald, with whom he had shared many of the vicissitudes of a professional Communist's life, including the war years in the Soviet Union. With the Communist seizure of power in February 1948, Slánský's own power also increased vastly, extending from the party secretariat into its control commission and cadre department, and via these into most organs of state security and public administration. He applied this power to prosecute the post-February mopping-up phase (see Chapter 3, section 3) so ruthlessly and terroristically that his unexpected disgrace in the late summer of 1951 was widely, but alas prematurely, hailed as signaling a return to moderation and lawfulness.

In addition to his apparent power, another reason why Slánský's fall seemed surprising at the time was that he had taken the lead during 1949 and 1950 in launching a purge of alleged Titoists, who included a number of senior military officers, the deputy minister of foreign trade, and—most fatefully—a cluster of senior Slovak Communists (including the minister of foreign affairs) who were pressing for the extension, or at least the retention, of Slovak administrative autonomy against Czech centralization pressures. In this affair, Slánský and Gottwald were still pulling together. But at the turn of the year 1950/51, direct Soviet instigation prompted Gottwald to order the arrest of a number of Slánský's protégés in the party apparatus. As 1951 was proving to be a difficult year economically, the temptation to deflect public discontent onto scapegoats waxed irresistibly, and Gottwald and Stalin now hit on the alluring formula of merging the former's temptation to this effect with the latter's burgeoning anti-Semitism and current obsession with his Titoist nemesis.

The upshot was a grand purge that cleared the Czechoslovak party leadership of all real, imagined, and potential rivals to Gottwald and skeptics about Stalin. Its most vividly targeted victims were that leadership's Jewish intellectuals and functionaries, Slovak autonomists, and wartime non-"Muscovites" ("undergrounders" as well as "Londoners"). At a spectacular show trial on November 20 to 27, 1952, marked by emphatically anti-Semitic rhetoric and ambience, Slánský and thirteen co-defendants—all seasoned veterans of the Communist apparatus—stood accused as "Trotskyist-Zionist-Titoist-bourgeois-nationalist traitors, spies, and saboteurs, enemies of the Czechoslovak nation, of its People's Democratic order, and of Socialism." Having been physically and psychologically "prepared" while under interrogation, they not only pleaded guilty, but also delivered incredibly self-abusing and self-abasing confessions, and several heard themselves denounced by wives and children.[7] Predictably, all were found guilty, three sentenced to life imprisonment, and eleven to hanging. In a macabre coda, the bodies of those hanged were cremated and their ashes spread as filler material into the ruts of a slippery cart road on the outskirts of Prague. Less spectacular Czechoslovak purge trials continued even after Stalin's death, in May 1953 and April 1954. Now the anti-Semitic tone was dropped, and their thrust was against only alleged "Titoism"and "Slovak separatism." Meanwhile, Gottwald had not been destined to enjoy his triumph over Slánský for long, dying on March 14, 1953, literally in the wake of his mentor, Stalin.

The Hungarian purge presented yet a third alternative scenario: the victors were "Muscovite" Jews; the vanquished were "local underground" Magyars; and the outcome was deadly. Ideological and policy issues were less marked than in Poland, but somewhat more so than in Czechoslovakia; on balance, the driving energy was a personal power struggle.

László Rajk, a Communist since 1931, a veteran of the Spanish Civil War, and the leading Hungarian "underground" Communist during World War II, had been arrested in November 1944, during the culminating paroxysm of German and native Radical Right rule that had followed and reversed Regent Miklós Horthy's bungled attempt to abandon the Axis and switch sides (see Chapter 2, section 4). After returning to Hungary from a German concentration camp in May 1945, Rajk was immediately appointed to the Communist party's Politburo, designated deputy secretary-general as well as head of the party's key Budapest organization, and usually listed as fifth in its hierarchy, immediately after the four recently returned Jewish "Muscovites"

Mátyás Rákosi, Ernő Gerő, Mihály Farkas, and József Révai. As interior minister from March 1946 until August 1948, Rajk orchestrated the destruction of the Smallholders, the absorption of the Social Democrats, and the tainted elections of August 31, 1947. He also used to encourage "toilers" spontaneously to punish "speculators" and "black marketeers," a stance that amounted to inciting and protecting occasional anti-Semitic excesses. In short, Rajk was a tough, ruthless, even fanatical Communist, not the kind of man to entertain principled objections to Stalinist methods. But he had pressed for a more rapid and total Communist seizure of power in 1945 to 1947 than had been authorized by the more cautious and methodical Stalin, who was still stretching out the coalition phase in Hungary and Czechoslovakia in order not to alarm the British and Americans prematurely. Thus the timing rather than the substance of Rajk's policy preferences placed him somewhat out of phase with Moscow's priorities and would soon expose him to charges of Titoism in the hysterical and sycophantic atmosphere that gripped the Communist world after the rupture between Tito and Stalin. Whereas Gomułka's heresies were nationalism and caution, Rajk's (if any) were internationalism and radicalism. His relegation to the less powerful foreign affairs portfolio in August 1948 marked the beginning of his decline, which then proceeded to his arrest in June 1949, spectacular show trial (together with a cluster of other wartime "undergrounders") in September, and hanging on October 15, 1949.

The trial of Rajk and his co-defendants was marked by the usual implausibility, mendacity, paranoia, and degradation of such affairs. The defendants confessed to prolonged treasonable service for the fascist–imperialist syndicate of Horthy-Himmler-Dulles-Tito. In addition to the prominent victims in the public dock, 2,000 Communist cadres were summarily executed, 150,000 imprisoned, and 350,000 expelled from the party—the victims being preponderantly wartime "local undergrounders," "Westerners," veterans of the Spanish Civil War, former Social Democrats, and senior military personnel. In effect, Rákosi killed more Communists in five years than Horthy had in twenty-five, and the entire society was cowed into a condition of bewilderment and terror.

It is something of a mystery how and why the Jewish quartet of Rákosi, Gerő, Farkas, and Révai were able to carry out this purge in the midst of Stalin's pronounced anti-Semitic obsession, which lasted from 1948 to 1953. Perhaps their wartime "Muscovite" status gave them some camouflage. More plausibly, their purge was preemptive, coming very early in Stalin's culminating crisis and eliminating any leadership alternative to themselves. If so, their selection of Rajk as the rival to be liquidated was astute, precisely because his radical, totalitarian propensities during the years 1945 to 1948 were about to be adopted by Stalin himself in reaction to the Titoist crisis, and hence their own survival required eliminating Rajk as a possible fulcrum for Stalin's new course.

In Romania, the first round of purges had preceded the Tito–Stalin rift. The local analogue to Gomułka, the wartime "undergrounder" Lucreţiu Pătrăşcanu, an ethnic Romanian and a relative moderate, had been politically (though not yet physically) eliminated in February 1948 by the "Muscovite" troika of Emil Bodnăraş, Vasile Luca, and Ana Pauker, all of whom were of non-Romanian ethnic-minority stock (see Chapter 3, section 6). But the fact that this episode preceded Tito's defiance of Stalin suggests that it had a different, albeit perhaps parallel, cause, either Pătrăşcanu's presumed dissent from Andrei Vyshinsky's heavy-handed intervention in early 1945 to impose the Communist-dominated Groza government on Romania or his known nationalistic proclivities. Then, in the spring of 1952, the "local" secretary-general of the party, Gheorghe Gheorghiu-Dej (an ethnic Romanian who had spent the decade 1933 to 1944 in royal jails), made a startling recovery from seeming political eclipse during the preceding winter by allying with Bodnăraş against the other two "Muscovites" as well as against the widely hated (albeit "local") Interior Minister Teohari Georgescu to establish his primacy in the party and to elevate himself to the premiership in place of Groza, whose "front" role was no longer needed.[8] As in the other people's democracies, this convulsion at the top was accompanied by a mass purge in the ranks of the recently merged Communist–Social Democratic Workers party. The intention was to render the Communist regime somewhat more indigenous, hence less unpopular, without substantially altering its policies and, indeed, while enhancing its centralized discipline. Simultaneously, Gheorghiu-Dej and his cohort were careful to articulate the required revilements of Tito and homages to the Soviet Union.

While the senior Bulgarian Communists Georgi Dimitrov and Vasil Kolarov had been making vivid careers abroad in international Comintern politics during the 1930s and 1940s, their party's underground apparatus at home was led by Traicho Kostov, a man of great courage and tenacity whose repeated subjection to arrest and torture by the police had left him partially crippled. With the installation of the Fatherland Front government in September 1944, Kostov had emerged as secretary-general of the Communist party, vice premier, and chief administrator of the national economy, which meant, in effect, the energizer of Bulgaria's early collectivization of agriculture and its planned industrialization and electrification. But though these were eminently respectable policies from a Stalinist perspective, Kostov was soon at loggerheads with Moscow because he was not shy about venting his objections to sharp and exploitative Soviet trade practices toward his country. He resented, for example, the Soviet Union's forced monopsonist purchase of Bulgaria's valuable rose-oil and tobacco crops for nonconvertible rubles and subsequent sale to the French perfume and cigarette industries at astronomic profits.

In the aftermath of the Tito–Stalin rift, such nationalism on the part of even a devout, proven veteran Communist would inevitably be judged a sinister heresy. On March 26, 1949, Kostov was dismissed from his party and governmental offices, and on June 25, was arrested, together with a number of senior Communist economic and military administrators. Though well "prepared" (like Slánský and Rajk) for the group show trial that opened on December 7, 1949, Kostov on the witness stand again demonstrated his exceptional toughness by repudiating the confession that had been extracted from him in the pretrial "softening-up" process—the by-now conventional confession to criminal espionage in the service of the prewar police and Trotskyists, then of the Western intelligence services and the Tito gang. (Ironically, during World War II, Kostov had been particularly energetic in contesting the Titoists' aspirations to Macedonia.) Promptly silenced and, parenthetically, mocked for the deformity received at the hands of the prewar police for his Communist militancy, Kostov was hanged on December 16, 1949. The Bulgarian Politburo took this occasion to cable its gratitude to Stalin, whose "deeply penetrating eye" had first discerned the treasonous infamy of Kostov and his clique.[9]

As Dimitrov had died on July 2, 1949, and as the elderly Kolarov would soon follow him on January 23, 1950, the immediate political beneficiary of Kostov's liquidation was to be Dimitrov's brother-in-law, the wartime "Muscovite" Vŭlko Chervenkov.

The Albanian consequences of the Tito–Stalin rift were described in Chapter 3, section 8.

Two general reservations ought to be borne in mind by way of summarizing these purges in the people's democracies: (1) although there obviously were winners and losers, even the winners remained insecure during the rest of Stalin's life; and (2) the purges were not just a lethal musical-chairs game among a limited number of players at the apex of the Communist elites. Rather, their demonology swelled and proliferated until it terrorized whole societies and snuffed out the residual creativity of entire political systems. Every inevitable failure to fulfill the unrealistic targets of the economic plans imposed by Stalin was henceforth imputed to saboteurs and spies. Meanwhile, the real causes of these failures remained unacknowledged and unexamined. Facts became taboo and suspicion pervasive. Politically, power drifted toward the security services and toward small, overly centralized, summit cabals, while the inner life of the Communist parties petrified. An orgy of denunciations and evasions of responsibility was unleashed throughout the upper and middle levels of the bureaucracy. As a consequence, the very idea of politics became degraded and execrated in the minds of the citizens, as it seemed to them synonymous with lawlessness, mendacity, and opportunism.

4

The Yugoslav response to these purges in neighboring states and to the Stalinist vendetta that they expressed was rather astute. Just as the Titoists had initially redoubled their Communist orthodoxy in the economic realm in order to prove their ideological rectitude and to enhance their revolutionary self-respect (see Chapter 3, section 5), so in the political area their first reaction to the anathema that had been pronounced against them was to admit to no deviation, to shun any revisionist experiments or concessions to anti-Communist forces either domestic or foreign, to press ahead with "building socialism" and combating imperialism, to refrain from criticizing Stalin or the Soviet Union, and instead to prosecute the quarrel by proxy—directing their counterthrusts at the Cominform apparatus and the satellite Communist parties. And even here they pretended that deteriorating interparty relations need not contaminate correct interstate relations. Thus the Fifth Congress of the Yugoslav Communist party, held at the end of July 1948, still featured rhythmically orchestrated chants of "Stalin, Tito, party" and portraits of both leaders. At the next month's International Danube River Conference in Belgrade, the Yugoslav hosts endorsed the Soviet proposal to exclude the nonriparian Western states from representation on that river's regulatory bodies, and at that autumn's meeting of the United Nations General Assembly, the Yugoslav delegation again systematically voted with the Soviet bloc. As late as May Day 1949, Stalin's portrait was still being carried icon-like along with those of Marx, Engels, Lenin, Mao, and Tito in the Yugoslav processions.

By then, however, these pretenses had worn thin—indeed, had become confusing to their Yugoslav audience—and were soon dropped. The trials of Rajk and Kostov in September and December 1949 prompted an extension of the quarrel from poisonous interparty relations to hostile interstate relations, as the Soviet Union, followed by all the people's democracies except Albania, used the trials as a pretext to denounce their treaties of friendship, mutual assistance, and cooperation with Yugoslavia and to slash their trade with it.[10] In its turn, Yugoslavia prohibited further use of its territory as a sanctuary by the Greek Communist guerrillas, thus effectively ensuring their defeat in the festering Greek Civil War and earning the approbation of the West, which proceeded to compensate Yugoslavia for its severed Eastern trade with loans, credits, and aid. Though grateful, Tito still insisted that Yugoslavia must regard itself as a Communist state, and he refused to make concessions to political pluralism or to market capitalism. It should be acknowledged, in this connection, that during this first year following their expulsion from Stalin's empire, the Yugoslavs were not only astute but also courageous, for they had no assurance of alternative support from a seemingly bleak and friendless world. Their political courage in this context replicated their military fortitude of the Partisan years.

By the winter of 1949/50, the Yugoslav leadership felt sufficiently confident of the support of its own party ranks and general public that it went over to the ideological counterattack. Instead of merely refuting Soviet and Cominform accusations, it claimed for itself the correct interpretation of Marxism-Leninism and accused the Soviet Union of revisionism, characterized by chauvinistic imperialism in its foreign policy and by bureaucratic despotism in its domestic structure. As evidence of their own authentically socialist transcendence of such degenerative processes, the Yugoslavs initiated in early 1950 their system of participatory workers' councils, to which many of the economic functions of the state bureaucracy were transferred, thereby supposedly commencing the process of the "withering away of the state" that Marxist theory postulates. The abandonment of collectivization of agriculture in 1953 was a parallel exercise in dismantling the state-commanded economy in favor of more autonomous, self-managing social forces. Of course, ideology was here supplemented and energized by the Tito regime's political need for domestic mass support and productive motivation, which also accounts for such complementary measures of these years as reducing the size of the bureaucracy, decentralizing the federal administration, diluting the ostentatious privileges of the elite, curbing the arbitrary power of the police, expanding personal freedoms, and seeking *modi vivendi* with the country's religious bodies. Many of these liberalizations were codified in the new constitution of January 13, 1953, itself a radical departure from the Soviet-imitating one of January 31, 1946.

Although this bold Yugoslav reassessment of the legitimacy of Soviet institutions and of Stalinist practices was not informed by the intense moral soul searching that would later in the decade suffuse the Polish and Hungarian quests for humane socialism and though it was always sponsored and controlled by the regime, whereas in these other two people's democracies the process was symptomatic of the decay of the regimes, the logic of dialectical reasoning within a Marxist perspective necessarily generated some ideologically awkward questions for the Yugoslavs: What would be the point of the "withering away of the state" if the political and administrative monopoly of the Communist party was not also attenuated? And how could a professedly socialist country accept massive economic and military aid from the allegedly imperialistic West without compromising itself? The second question became imperative with the outbreak of the Korean War in June 1950; the first eventually spawned the Djilas crisis of January 1954.

Tito appears to have initially hoped that succor from Mao, freshly triumphant in the Chinese Civil War and impeccably revolutionary, would spare Yugoslavia the ideological embarrassment of heavy economic and military dependence on the West. When this hope proved vain, the Yugoslavs took up a flirtation with lapsed Trotskyists, expelled but still nostalgically leftist ex-Communists, Sartrian intellectual mandarins, American Wallaceites, self-styled progressives, and other *soi-disant* Leftists. But they proved to be derisively weak and, in any event, became altogether useless as they endorsed, excused, or ignored the Communist North Korean invasion of South Korea in mid-1950. For Yugoslavia, this invasion was a moment of truth, for if Soviet-supported aggression were to go unresisted and thereby be allowed to succeed in the Far East today, why not also in the Balkans tomorrow? Accordingly, Yugoslavia not only supported the United Nations' American-led collective-security response to this invasion (though not General Douglas MacArthur's later counterinvasion of North Korea), but went on to accept American reequipment of its own motorized and armored divisions as well as massive economic aid from the United States and several Western European countries. The ideological rationale was that these countries were no longer flatly bourgeois-imperialist, but were impregnated with the seeds of a peaceful and possibly early transition to socialism thanks to their free political institutions in general and to their vibrant Social Democratic parties in particular. These Social Democratic parties enjoyed five virtues in Titoist eyes: (1) most of them were genealogically and still sentimentally Marxist; (2) they were either currently or potentially governing parties in their respective states; (3) they were unimpeachably anti-Stalinist; (4) they were not neutral toward the Korean War; (5) they were responsive to the Yugoslav predicament.

As for the logical implications of the "withering away of the state"—that is, domestic structural and political transformations—most of the Yugoslav leaders trod more warily, though the dynamics of this issue eventually propelled them into a crisis from which ensued personal as well as political rifts, which were all the more painful for the warm Partisan-steeled comradeship that had preceded them. The liberalizing policies had been adopted without dissent, for they were needed by the regime to re-solidarize itself with its home public at a time of external peril. So also the economic decentralization that encouraged popular participation and was intended to give Yugoslavia a modified open-market economy without private entrepreneurs. But when the Communist party changed its name to the League of Communists in November 1952 (followed in February 1953 by the People's Front restyling itself as the Socialist Alliance of Working People) and instructed its members henceforth to lead by persuasion and education rather than to rule through "administrative methods," Tito found it necessary to reassure its baffled cadres that "Communists must still be on top" and remain vigilant in order to lead the country to full socialism. This pronouncement hardly sufficed to resolve widespread bewilderment and even demoralization, which were further compounded when Tito sponsored double-barreled condemnations in June 1953 by the League's Central Committee of both "bureaucratic" and "Westernizing" deviations—the latter rebuke couched in sharper language.

At this point, Milovan Djilas, one of the trio of top leaders who during and since the war had been closest to Tito, a Montenegran intellectual prone to follow and to drive the logic of ideas to their ultimate conclusion, formerly an inordinate admirer of Soviet models and extreme Stalinist, currently secretary of the League's Central Committee, president of the state's National Assembly and vice president of the Federal Executive Council, the regime's leading propagandist, published a series of articles arguing that the Leninist type of "vanguard" party had exhausted its historical role in Yugoslavia—indeed, had become a brake on further progress toward socialism—and therefore should make way for unfettered social spontaneity by the masses—that is, should wither away along with the state.

Even in the absence of external developments, it is unlikely that his comrades would have accepted Djilas's prescription, for they were realistically skeptical of the masses' "spontaneous" commitment to socialism. All the more so at this particular juncture, when the recent death of Stalin, on March 5, 1953, soon followed by his heirs' first conciliatory feelers, rendered Tito and the others reluctant to widen further the political, ideological, and structural gulf between Yugoslavia and the Soviet Union and, indeed, aroused these leaders' hopes of a *modus vivendi* based on a Leninist common denominator. But Djilas pressed on, rejecting Leninism as well as Stalinism, arguing that the former inevitably spawned the latter, and insisting that the "socialist consciousness" of the masses be trusted and that the League of Communists liquidate its apparatus and transform itself into a nonbureaucratic gathering of ideologically kindred colleagues. He supplemented this generalized political argument with a transparent *nouvelette à clef* vitriolically satirizing the personal morals of the country's highest functionaries and their wives as those of a greedy, snobby, coarse, parvenu caste—now sadly degenerated from the selfless warrior-peasant heroes they had once been. Though provoking predictable outrage in the stratum that saw itself as bearing primary responsibility for nurturing and protecting the Yugoslav state through critical times, Djilas's diatribe still aroused widespread confusion because it had come from deep within this magic circle—confusion that now required Tito's personal intervention at a special public plenum of the League's Central Committee, on January 16 and 17, 1954.

Speaking more in sorrow than in anger at an occasion where most of the other participants reviled and mocked Djilas, Tito criticized his lack of discipline, insisted on the continuing need for Leninist vigilance against foreign and domestic enemies of the system, and deprecated "anarchistic" speculation about the withering away of the League (party) as inexpedient and premature. Djilas was expelled from the Central Committee; in May, he resigned from the League after more than two decades of membership. He was also deprived of all his state offices, including the legislative seat to which his Montenegran constituency had reelected him as recently as November 22, 1953, by a 99.9 percent vote, and from which it now recalled him—an ironic comment on the level of spontaneous democracy in Titoist Yugoslavia. Though periodically harassed and occasionally jailed, Djilas was neither executed nor charged with treason; he went on to become a fearless literary, historical, and ideological critic of the Yugoslav regime and of international Communism in general.

The political lessons of this episode were: (1) though he had not been particularly active in the preceding doctrinal innovations, Tito was shown to be still the country's and the League's master because his intervention was required to resolve this issue; (2) Western ideas, though officially repudiated, were making inroads into Yugoslavia; (3) the League was having difficulties finding an ideological (albeit not an operational) middle ground between Stalinism and democracy; (4) for all their rhetoric about local initiative through the communes and the workers' councils, Tito and his entourage retained the classic Leninist mistrust of the political spontaneity of the masses. Hence their Yugoslavia settled for the next decade into a politically interesting and precarious balance between a relatively relaxed, experimental, nontotalitarian state and society on the one hand, and a monopolistic, authoritarian, potentially totalitarian, Communist party (League) on the other.

5

The Soviet-controlled people's democracies were meanwhile in the grip of mature Stalinism. It was characterized by the enforced imitation of Soviet political, administrative, and cultural institutions; absolute obedience to Soviet directives and even hints; administrative supervision by Soviet personnel; bureaucratic arbitrariness; police terror uncontrolled even by the local party; economic deprivation while pursuing overambitious industrial investment programs and undercapitalized agricultural collectivization drives ("lunar economics"); colonial-like foreign-trade dependence on the Soviet Union; isolation from the non-Communist world and to some extent even from other people's democracies; synthetic Russomania; a mindless cult of Stalin adulation; and resultant widespread social anomie, intellectual stagnation, and ideological sterility.

The economic, cultural, and moral costs of this mature Stalinism in the people's democracies are incalculable and were paid by millions of displaced, disenfranchised, dispossessed, dispersed, humiliated, and persecuted people. Politically, the system itself would pay a severe and ironic cost in the coinage of the alienation of its emblematically favored and supposedly ruling class, the industrial workers, who deeply resented the lack of tangible rewards for their hard work, the incompetence of their party-designated managers, the technological obsolescence of much of their new industrial equipment, and the distortions inflicted on their national economies and cultures. Their bill would be presented soon after Stalin's death.

5

The Revenge of the Repressed: East Central Europe Reasserts Itself

1

Tito's repudiation of Djilas's notion of the withering away of the Communist party was a prerequisite for the partial reconciliation between the Soviet Union and Yugoslavia that Nikita S. Khrushchev initiated after Stalin's death on March 5, 1953. Though the Yugoslav leaders were at first reserved toward the Soviet overtures, presumably because they were awaiting institutional and policy changes to give substance to the personnel rotations that were taking place in Moscow at that time, they eventually acceded to Khrushchev's wooing, albeit heavily on their own terms. After all, this partial reconciliation not only raised Yugoslavia's leverage internationally, but also enhanced Tito's standing with his own Communist cadres domestically. All their Partisan self-confidence and national pride at facing down Great Power pressure notwithstanding, these cadres had never been enchanted with the discipline-shaking and privilege-eroding Titoist innovations or fully persuaded of their Leninist orthodoxy. Hence the reconciliation, in which Moscow in effect restored the aura of ideological respectability to Belgrade, did much to revive their morale, on which Tito ultimately depended.

Soviet–Yugoslav diplomatic relations were rectified steadily during 1953 and 1954. Redressing reciprocal grievances over Danube River issues provided a simultaneous auxiliary route to remedying Yugoslavia's relations with the riparian people's democracies. But all these were still only alleviations of interstate relations and left unresolved the critical question of ideological responsibility for the Tito–Stalin rift. The Yugoslavs, determined not to be pressed into the posture of penitents, insisted that the Soviets acknowledge sole responsibility. Such an acknowledgment would, of course, transform the essence of international Communism, morally dethroning its Soviet overlord while exonerating and rehabilitating the Titoist schismatics and even yielding to them the ideological high ground. Hence Vyacheslav M. Molotov, who had cosigned with Stalin the Soviet accusations of 1948, objected strenuously, but Khrushchev—perhaps not fully appreciating the issue's implications—eventually finessed the Yugoslavs' condition by blaming the past troubles on Stalin's secret-police henchmen Lavrenti P. Beria and Viktor S. Abakumov, both now safely purged and executed as "enemies of the people." Though crude and transparent, this alibi sufficed to enable Khrushchev and other top Soviet leaders to be received by Tito in Belgrade on May 26, 1955. Yet the final declaration issued when they left a week later on June 2 was signed for the Soviet delegation not by Communist party leader Khrushchev but by the titular head of the government, Nikolai A. Bulganin, a signal that full ideological reconciliation was still beyond reach and a symbol of the Yugoslav contention that relations between two equal, sovereign states, not between a central and a peripheral party, were being mended. Tito signed for Yugoslavia in his capacity as the country's president.

Though each side now implicitly conceded that the other was indeed "socialist" and not "degenerate" or "fascist," as their previous polemics had charged, and though the declaration provided for the future "exchange of socialist experiences" between them, the Yugoslavs insisted on their by-now traditional point: "questions of internal structure, differences of social systems, and differences of concrete forms in developing socialism are exclusively a matter for the peoples of the different countries." They also inserted a phrase condemning military blocs, even though the Soviets had just formalized the Warsaw Treaty Organization (WTO) on May 14. The declaration omitted all allusions to Marxism-Leninism, to the Cominform, or to responsibility for the 1948 rift. On balance, Tito had maintained his independence and asserted his parity, while Khrushchev left with less than he had hoped for but enough to press on with his innovative policies of international coexistence, domestic "thaw," and reconstruction of the Soviet bloc.

What Khrushchev hoped to achieve through reconciliation with Tito seems clear. To be the restorer of the "unity of the socialist camp" would enhance his prestige domestically, in the people's democracies, and in China, at a time when his primacy among Stalin's heirs was still nascent and precarious and in a context where the chief opponents to his programs were the Stalinist hard-liners. The West, which for the past six years had been supportive and generous toward Yugoslavia, would look gullible and helpless. These prizes were sufficiently enticing to encourage Khrushchev to press on with his efforts to harness Yugoslavia's national Communism despite his hosts' relative coolness at their first encounter in Belgrade. Thus he again offered doctrinal as well as historical carrots the following February at the pivotal Twentieth Congress of the Soviet Communist party and once more during Tito's triumphal return visit to Moscow in June 1956, when *Pravda* unprecedentedly opened its pages to an ideological defense of Titoism by Vice President Edvard Kardelj of Yugoslavia. The long-moribund Cominform was dissolved in April 1956 after slavishly rescinding most of the anti-Tito condemnations that it had slavishly issued during 1948 and 1949.

But nemesis soon struck. Whereas the Soviet Union's own political apparat, though disconcerted, was sufficiently mature and resilient to absorb these dizzying cartwheels, the newer, more fragile ones of several people's democracies broke under the strain, which was compounded by Tito's persistently demanding his pound of flesh from the "little Stalins" of East Central Europe whose local power was anchored in their anti-Titoist purges of the recent past. There thus emerged the ironic spectacle of the Yugoslavs, who in principle championed national independence and separate roads to socialism, demanding that the people's democracies imitate the Soviet Union's current repudiation of Stalinism, an imitation that would jeopardize the very survival of their regimes at a perilous historical juncture. Their monopolization of political power and their forced industrialization and collectivization drives had been too recent and were still too brittle to have been organically consolidated. Hence their regimes, which symbolized total coordination with the Soviet model, now (also ironically) sought relief from the Titoist pressure (which they could no longer dismiss as simply heretical and deviationist) by pleading special national considerations. Khrushchev was thus caught in a triadic dilemma among his perceived need to reintegrate Yugoslavia into the Communist world, his craving to dismantle political Stalinism, and the Soviet Union's stake in averting the destabilization and possible loss of its client states.

2

Though there had been some rioting in Bulgaria, Czechoslovakia, and East Germany in May and June 1953 shortly after Stalin's death, these outbreaks had been provoked by nitty-gritty grievances over work conditions, monetary reforms, and food rationing; they had not involved the kind of systemic and ideological challenges to the Stalinist structural legacy that were educed in 1956 by the combination of Khrushchev's programs and Tito's exactions. The first people's democratic regime to succumb to this strain was Poland's.

Our earlier narrative had left the relatively "national Communist" former Secretary-General Władysław Gomułka neutralized, ostracized, and arrested by the Polish "Muscovites" between 1948 and 1951, followed by the conventional imposition of mature Stalinism (see Chapter 4, sections 3 and 5). In one dimension, this mature Stalinism was less rigorous in Poland than elsewhere; in another, more so: agriculture in Poland was less collectivized than in the other people's democracies, but Poland was obliged to assume a heavier share of the conversion to military production that Stalin imposed on his bloc after the outbreak of the Korean War.[1]

The Polish regime's first reaction to Stalin's death was to concede to the public some economic alleviation but to refuse any political relief. This potentially clever strategy was aborted when the execution of Beria, the veteran Soviet secret-police chief, in December 1953 panicked the associated Polish secret-police apparatus,one of whose top officials now fled to the West, from where his detailed revelations of terror, corruption, depravity, and opportunism in the Polish elite began to be broadcast back into Poland from September 1954. The effect of these revelations was devastating; even the party cadres were appalled. The police apparatus was thereupon sharply curbed and purged, while Gomułka was quietly released from detention in December 1954. The next month, the Central Committee bitterly criticized its own Politburo for having permitted such excesses.

The Soviet leaders' rapprochement with Tito in May 1955, followed by the Geneva Big Four summit meeting in July and the Warsaw international youth festival in August, accelerated Poland's domestic political erosion. They rendered the outside world more familiar and respectable and exposed the grim decade of postwar Polish developments as a tawdry and gratuitous betrayal of bright promises. Party-supervised youth and cultural, intellectual, and press circles now became restive—before any worker turmoil—and rebuffed senior party efforts to discipline them. Khrushchev's denunciation of Stalin at the Twentieth Congress of the Soviet Communist party in February 1956 released a torrent of soul searching in Poland. The parliament ceased to be a rubber stamp; thousands of political prisoners were amnestied, and several of the most brutal police officials were arrested; the special stores catering to the privileged were reduced, obliging the elite to rub shoulders with the citizenry; the wartime London-loyal resistance movement was partially rehabilitated; and Gomułka's earlier secret release was now publicly acknowledged, but as yet without his political vindication—he was still officially deemed a deviationist, though no longer an enemy. By and large, the thrust of Polish society's moral and ideological groping in the spring of 1956 was not so much for an anti-Communist system as toward a combination of humane Marxism with nationalistic idealism, for "sincerity," implemented through real policy reforms. That is why the issue of Gomułka, of whom people had been whispering ever since 1948 that "he defends Poland while others sell out Poland,"[2] was so important.

The death of the veteran Stalinist party chief Bolesław Bierut on March 12, 1956, had ambivalent consequences. On the one hand, by opening a leadership gap, it fanned the current commotion; on the other hand, it spared Poland a wrenching struggle to remove him, such as was lacerating Hungary over Mátyás Rákosi at this time (see section 3). Bierut's successor as Communist party first secretary, Edward Ochab, sought a stance of balance and gradualism, further leashing the police, easing the inordinate investments in heavy industry, relaxing the pressure on Catholic believers, while condemning Gomułka for nationalism and insisting on preserving the Communists' monopoly of power. Hence he was deemed acceptable by the Soviets, whose own domestic de-Stalinization and external rapprochement with Tito would have been compromised had they interceded in Poland on behalf of their erstwhile (and still sentimentally favored) clients and policies.

But Ochab's straddling became untenable after the Poznań workers' insurrection of June 28 and 29, which indicated that the previous ferment could no longer be contained within the existing party structure—albeit now a self-criticizing structure—or confined to the intelligentsia. The workers rioted to protest shortages of food and consumer goods, bad housing, declining real income, shipments of commodities to the Soviet Union, inept management, waste, and bureaucratism. Scores were killed and wounded. Even more ominously, the army was rather demoralized by being used to suppress the insurrection.

Ochab's new administration, which originally and expectedly blamed this affair on "provocateurs, counterrevolutionaries, and imperialist agents," then strikingly broke step with Moscow by shifting to the interpretation that the Poznań rebels were honest workers driven to desperation by valid grievances. And it decided not to be paralyzed into halting its liberalizing trend but to accelerate it, despite the profound reservations of its still strong "Muscovite" contingent, currently named the Natolin group (after an estate near Warsaw where it caucused). Thus during the summer of 1956, many economic failures were exposed and several senior Stalinist planners and administrators were consequently removed. Nevertheless, workers were spontaneously electing workers' councils independently of both the party and the official trade unions—a vivid signal that administrative changes without political reforms would not suffice. Simultaneously, the pilgrimage on August 25 and 26 of well over 1 million people to the Jasna Góra monastery at Częstochowa to mark the three hundredth anniversary of this shrine's supposedly miraculous deliverance from a foreign army's siege was a combined national and religious intimation that among the political reforms that the public was expecting would be a reassessment of Polish–Soviet relations.

The Natolinites proposed to finesse this pressure for political reforms with a quartet of demagogic recommendations: (1) co-opting Gomułka to the current Politburo; (2) releasing Stefan Cardinal Wyszyński from his monastic internment; (3) raising wages by 50 percent; and (4) purging a number of prominent Jews from public life. But Gomułka rebuffed the Natolinites, refusing to return to office without the power to set policy, insisting that collectivization of agriculture he abandoned as a failure, demanding that Polish national interests be asserted even vis-à-vis the Soviet Union, and requiring a repudiation of the official rationale for his purge in 1948 to 1951 and a reversal of its outcome.

At this point, Ochab showed rare statesmanship by deciding to yield the first secretaryship of the party to Gomułka and to let him reconstitute the Politburo, as the only hope for satisfying the masses while saving Polish Communism. This despite the fact that Gomułka clearly intended to drop the Soviet Marshal Konstantin K. Rokossovsky from the Polish Politburo and Defense Ministry as a signal of restored national dignity. The Soviets feared that this scenario would jeopardize both their own hegemony and Communist power as such. But during a tense and bitter political confrontation in mid-October 1956, they ascertained that the Polish army and workers would fight to defend Poland's right to determine the composition of its own leadership (a Titoist claim) and ruefully accepted these changes. The self-effacement of Ochab left the Soviets with no credible challenger to Gomułka, for they knew that the Natolinites had little local support, that Mao Tse-tung approved Gomułka's elevation, that the Stalinist legacy in Poland's domestic and international situation needed correcting, and that using force against Poland at this point would undermine the reconciliation with Tito on which Khrushchev was banking so heavily and would succor his own residual Stalinist rivals.

In return for Soviet tolerance, Gomułka committed himself to preserving the Communist monopoly of power—albeit in a more national form—and to holding Poland to its Soviet alliance—albeit on more equal terms. This commitment was not extracted from him but expressed his own values, unchanged from 1948. In turn, Khrushchev's acceptance of its nuances was compatible with his own de-Stalinization course and his groping for a less monolithic yet still reliable bloc. In now fulfilling his commitment, Gomułka was to be vicariously aided by the spectacle of the crushing of the Hungarian Revolution by Soviet troops in November, which served as a sobering lesson to the still restless and elated Polish people of the dire consequences likely to ensue were they to press their significant yet limited political success too far, toward dismantling Communist power and seceding from the Soviet bloc. "Hungary" taught the Poles that Gomułka's national Communism was the maximum achievable at that time, and the newly released Cardinal Wyszyński now applied his moral authority to help teach this pragmatic lesson.[3]

3

Though the Communist monopolization of power had been imposed on Hungary considerably later than on Poland (see Chapter 3, section 4), its subsequent telescoping into a condensed, intensive period had rendered it especially traumatic. In Hungary, the cult of Stalin was particularly egregious, the treatment of "undesirable" social categories (for example, the bourgeoisie) particularly cruel, the superindustrialization commitment particularly heedless, and the circle of decision makers particularly restricted and "Muscovite." Presiding over this orgy of political sycophancy and economic irrationality was the arch-Stalinist Mátyás Rákosi, on whom Tito was particularly determined to wreak vengeance (see Chapter 4, section 3).

After Stalin's death, his Soviet epigones opted for collective leadership and repudiated the so-called cult of the individual. Given the still prevailing atmosphere of mandatory emulation of the Soviet model, this meant that the "little Stalins" in the people's democracies were similarly obliged to divest themselves of some of their plurality of offices and to collectivize their countries' leaderships. At a minimum, they could no longer combine in their own persons the first (or general) secretaryship of their respective Communist parties with the premiership of their governments, just as Khrushchev and Georgi M. Malenkov had separated these two offices in the Soviet Union. Accordingly, Rákosi—like his Polish, Czechoslovak, Albanian, and East German counterparts—retained the party post while reluctantly yielding the governmental one. In Romania and Bulgaria, on the contrary, the local strongmen reversed this pattern, imitating Malenkov.

The new Hungarian premier, as of July 4, 1953, was the wartime "Muscovite" but non-Jewish and uncommonly moderate Communist Imre Nagy (not to be confused with the Smallholder premier of 1946 and 1947, Ferenc Nagy; see Chapter 3, section 4). He had manifested his moderation most vividly in the late 1940s by his skepticism toward Rákosi's drive to collectivize agriculture rapidly and coercively, which earned him moral credit among the peasants. Though punitively suspended at that time from his party functions, Nagy had been spared the lethal fate of Rajk thanks to Soviet protectors (Beria and/or Malenkov?). Now, with the economy approaching collapse and hunger stalking this traditionally food-exporting country, he was installed in the Hungarian premiership at Soviet behest with the assignment of bridging the chasm between the regime and the people, especially the peasants, by turning the economy and the polity toward the so-called New Course—promoting living standards, consumption, wages, decollectivization, agricultural and light-industrial investments, artisanal private enterprise, administrative decentralization, popular participation, educational liberalization, lawfulness, amnesty, and religious toleration. This program was the sharpest and earliest reversal of mature Stalinism to be initiated in any people's democracy, and its very abruptness would soon entail politically costly disorientation and demoralization.

Though Rákosi's erstwhile policies were now denounced as a "megalomanic swindle," he was permitted to retain control of the party apparatus through its Secretariat, and he used that control to sabotage Nagy's reforms. Thus the interim standoff between Malenkov and Khrushchev in the Soviet Union allowed a fatal split between state and party to open in Hungary, leaving the country under a diarchic rather than a collective leadership, stalemated between Nagy's program and Rákosi's apparatus. Moreover, Rákosi's deliberate sabotage of Nagy's New Course was compounded—presumably inadvertently—by a Soviet foreign-trade retrenchment in 1955, which reduced the supply of raw materials and capital goods to Hungary just when its revised economic priorities required more goods and fresh credits. Finally, the ouster of Malenkov in February 1955 eliminated that Soviet leader who was most protective of the New Course, and Nagy fell in his wake in April—ousted from the premiership, from his parliamentary seat, from the party's Politburo and Central Committee, and even from his residual university lectureship and membership in the Academy of Sciences. In November, he would be expelled from the party altogether.

Nevertheless, Rákosi's resurgence proved pyrrhic. While the freshly ascendant Khrushchev indeed endorsed a reversion to neo-Stalinist economic priorities (for example, heavy industry and agricultural collectivization), he vetoed any return to the politico-administrative terror that would have been required to energize such a program in a Hungary on which the recent Nagy interlude had made a profound moral and psychological impact, exposing to masses and cadres alike the brittle and depraved nature of Rákosi's system.[4] And in foreign policy, too, Khrushchev's determination to court Tito gravely jeopardized Rákosi. Thus the abrupt termination of Nagy's experiment in April 1955—as abrupt as its initiation in July 1953—failed to resolve the crucial issues of Hungarian power and policies and left the country in continuing political suspension. Both Nagy's reforms and Rákosi's restoration were half-measures, and it has been a truism of politics since Machiavelli that half-measures are unstable and thus dangerous. Though Nagy was now once again officially a "nonperson" and though he had throughout been politically naïve and unskillful, his stature stood high, and the memory of him as an alternative ticked like a time bomb under the resurgent Rákosi.

Rákosi now fell between two stools. On the one hand, his announcement of a principled return to Stalinist priorities aroused dismay; on the other hand, his inability to enforce their implementation elicited contempt. Initially, the erosion of his system and his authority proceeded more rapidly, more vocally, and more dangeorusly at the elite level—within the party and especially among the intelligentsia—than at the mass level. Hungarian intellectuals, especially writers, have a tradition of regarding themselves as the polestar and the conscience of the nation, and they resented the absurdity of their being expected first to celebrate

Stalinism until 1953, then to hail the New Course, and now to applaud its repudiation. Unlike apparatchiks, they were not habituated to viewing their own past stances as irrelevant whenever the official line might change. Thus the progressive and Marxist writers, who had naïvely and sincerely vilified Rákosi's purge victims at the time of their arrests and trials in the late 1940s, had been deeply mortified when some of the surviving ones were released and told their tales of terror in the summer of 1954. In shame and remorse, the writers thereupon resolved never again to lie and never again to serve an inhumane purpose, no matter how "historically necessary" it purported to be. They now stood their ground against Rákosi, repelling his Central Comittee's efforts to violate their union's organizational autonomy, outflanking his demagogic attempt to mobilize the latent anti-intellectualism of the workers against them, and exploiting the protective cover of Khrushchev's denunciation of Stalin's excesses in February 1956 to oblige Rákosi to rehabilitate László Rajk posthumously a month later.

But even this major concession only compounded the problems of the beleaguered Rákosi, for his hypocritical insinuation that the purge of Rajk in 1949 had been a police provocation that had entrapped the party leadership (that is, himself) into falsely suspecting an innocent comrade (that is, Rajk) enraged the security services and offended the party cadres. Nor did his amnesty of some Catholic prelates and the release of several thousand more political prisoners help.[5] Though he was by now clearly a liability to the Communist system and to the Soviet stake in Hungary, Rákosi refused to resign lest (as he shrewdly argued) the entire political edifice crumble with his departure. Indeed, in mid-July 1956, he boldly pivoted from this recent string of concessions back to his customary preference for the mailed fist by proposing to arrest Imre Nagy and several hundred intellectuals and to shut several journals. Coming on the heels of the Poznań workers' insurrection in Poland, whose contagious potential was widely recognized, this prescription struck both the Soviet leadership and most members of Rákosi's own Central Committee as incendiary. Accordingly, on July 17, Soviet Presidium member Anastas I. Mikoyan went to Budapest and, assisted telephonically by Khrushchev, coaxed Rákosi into resigning as first secretary for reasons of hypertension (a plausible alibi under the circumstances) and retiring to the Soviet Union (whence he never returned and where he died on February 5, 1971, still professing the correctness of his positions).

In selecting Rákosi's successor, the Soviets again blundered into half-measures. Rather than Nagy, whose designation might indeed have calmed the gathering political storm, they chose Ernő Gerő (see Chapter 4, section 3), who was too closely identified with Rákosi and his excesses to elicit either the confidence of the elites or the trust of the masses. Mikoyan's maneuver thus entailed a number of liabilities: it demoralized the residual Hungarian Stalinists without appeasing either the reformers or the still implacable Tito (until October, when Tito belatedly relented toward Gerő); it was manifestly a Soviet rather than a Hungarian initiative; and it whetted the appetite for real political change without satisfying it, and thereby probably catalyzed the Hungarian Revolution.

While Ochab in Poland was wisely drawing Gomułka back into the political process during the summer and autumn of 1956, Gerő imprudently continued to ostracize Nagy. (Only under Soviet and Yugoslav goading was Nagy restored to party membership, but not to any office, on October 13.) Seeking to placate the intellectuals, Gerő threw several fellow Stalinists to the wolves and staged a macabre ceremonial reinterment of Rajk's bones—in vain. Fearful lest rapid reforms precipitate an avalanche, he moved too slowly. In sum, the Hungarian party leadership, unlike the Polish one, failed to dissociate itself convincingly from Stalinism or to identify itself as authentically patriotic.

So far, the Hungarian revolutionary process had been largely an internal elite process; the workers and peasants, less interested in reforming than in abolishing the Communist system, had remained sullen and silent. The spark that activated their mass participation was the reluctant Soviet acceptance of Gomułka's return to power in Poland, which the exhilarated Hungarians misinterpreted as a general Soviet abdication in East Central Europe; the linkage between the elite and the mass processes in Hungary was provided by the students, who were organizationally within the Communist system but outside its power structure.[6]

On October 23, an initially peaceful but politically militant demonstration by Budapest students at the statue of a nineteenth-century military hero of both the Polish and the Hungarian independence struggles—a demonstration intended to express solidarity with the contemporaneous political devolution in Poland and to urge its replication in Hungary—precipitated the revolution. The students were demanding the return of Nagy to power; the punishment of the violators of legality during the Stalin–Rákosi years; the evacuation of Soviet troops together with a reappraisal of Hungarian–Soviet relations; freedom of speech, press, and the like; the reorganization of the economy, including the revision of work norms in industry and the easing of agricultural delivery quotas; the rights to strike and to vote in free, multiparty elections; the release of all political prisoners; the readoption of the pre-Communist national coat of arms; and the removal of a statue of Stalin. The workers and general population quickly took to the streets in solidarity with the students; the mass process now swamped the elite process. Shooting erupted between the demonstrators and the security police, while the army remained neutral—in effect, benevolently neutral toward the crowds. Gerő, conceding that Nagy might again become premier, insisted on retaining the party first secretaryship, denounced the students' (actually by now the general population's) political demands as impudent, chauvinistic slander, and appealed for Soviet intervention and rescue. (Once again, his behavior was the opposite of Ochab's in a somewhat analogous situation.) The next day, October 24, the Soviet troop garrisons in Hungary, operating under the erroneous impression that the workers would not support the students, did attempt to suppress the revolution, but on discovering the error of their assumption, soon halted their intervention in the provinces and by October 28 in the capital as well. Politically, too, the Soviets were reassessing their commitments, and on October 25 brought about the replacement of Gerő as party first secretary by János Kádár. A wartime "local undergrounder," a tortured victim of one of Rákosi's last purges, and non-Jewish, Kádár was, in short, Gerő's opposite and symbolic refutation. The Soviets also endorsed Nagy as premier and anticipated that the Nagy–Kádár team would now master the situation, presumably toward an outcome comparable with the one that had been negotiated in Poland.

But Nagy not only was trapped in an objectively most difficult situation, with his public overconfident, his party in shambles, his army confused, but also, during his most recent ostracism, had subjectively moved quite a distance from his moderate reformist New Course position of 1953 to 1955. He now no longer believed in a necessary, "enlightened" congruence of Hungarian national and Soviet Great Power interests within a Soviet-dominated bloc; he inferred that the status of a satellite would always compromise efforts to build socialism in Hungary and hence that effective national independence was a precondition for authentic socialism; but since true small-state sovereignty was not feasible in a bipolar world of competing blocs (NATO and WTO), he concluded that the blocs should be dissolved.[7] This incipient intellectual and political neutralism hardly rendered Nagy a plausible candidate for the role of "Hungarian Gomułka," quite apart from his tactical political weaknesses and the difficult revolutionary situation into which he was thrust.

Since the first Soviet military intervention had been inconclusive, the Hungarian revolutionaries erroneously imputed to Moscow a general loss of political will. And since that intervention had been formally fig-leafed by the WTO, they demanded that Hungary denounce and leave this pact. And since the Communist party not only was perceived as an antinational entity, but also had virtually disintegrated, they demanded free multiparty elections. First articulated by the workers of the northwestern town of Győr—who, like workers throughout the country, were now acting through spontaneously created new workers' councils rather than through the discredited official trade unions—these twin demands were adopted well-nigh universally by a people that, on the one hand, had been embittered by the initial Soviet military intervention and, on the other hand, was misinterpreting the intermediate phase of Soviet temporizing.

The Nagy–Kádár team desperately sought to satisfy both the Soviet expectation that it restore an acceptable political order and the impulses of the revolution. In the end, it was torn apart by these incompatible pressures, with Nagy eventually opting for the latter and Kádár for the former. Some released Smallholder politicians quickly joined the government, but the old Social Democrats initially declined to do so unless their party was restored as an independent body. Given the current disintegration of the Communists, Kádár and the Soviets deemed this condition unacceptable, but Nagy acceded to it on October 30. József Cardinal Mindszenty was freed that same day, but (in contrast to the Polish primate's helpful stance toward Gomułka) declined to support Nagy. Also on October 30, Moscow announced its readiness to reexamine the institutional structures of its relations with the people's democracies and to abide by any unanimous request of the WTO members that it withdraw its garrisons from the territory of any one of them. But the first part of this Soviet declaration was too late and the second part too transparent to conciliate the revolutionaries. Over the next two days, Nagy threw his lot in with them and burned his bridges by (1) repudiating Gerő's original request of October 23 for Soviet military intervention; (2) reconstituting the multiparty government coalition of 1945; (3) withdrawing Hungary from the WTO; (4) proclaiming its international neutrality and requesting the protection of the United Nations.

To the Soviet leaders, this was triply unacceptable. They perceived it as an affront to their Great Power status, as upsetting the balance of power in Europe and even globally, and as sociohistorically retrogressive. Having already sent fresh, "uncontaminated" troops into Hungary since the early hours of November 1, even before Nagy had finalized his third and fourth fateful steps, they now crushed the revolution, starting on November 4.[8] In the wake of their armies, Kádár painstakingly rebuilt Communist power, suppressing both "revisionist" (Nagy-supporting) and "dogmatist" (Rákosi-nostalgic) competitors. Nagy was tried and was executed on June 16, 1958. Logically, albeit shamefully, he was in the end also abandoned by his earlier champion, Tito, in whose judgment Nagy had proved to be not a strong national Communist, but a hapless dabbler in the unacceptable, un-Leninist, withering-away-of-the-party reveries of Djilas. Of course, Tito was also one of the losers in the Hungarian Revolution, for his inability to "deliver" a peaceful transition there, either through his belated embrace of Gerő or through his sponsorship of Nagy, hurt his standing in Moscow and soon contributed to a second dispute between the Soviet Union and Yugoslavia.

It should be noted, by way of a coda, that these events in Hungary in 1956 were not a mere rebellion or revolt or uprising or insurrection or *Putsch* or general strike, but a genuine and domestically victorious revolution with national-political as well as socioeconomic aims. This revolution was defeated only by overwhelming foreign force, which by invading implicitly conceded the hollowness of the vaunted social pillars of Communism—the workers, peasants, intelligentsia, youth, and cadres. Comparing the Hungarian Revolution with the almost simultaneous events in Poland taught that, while Stalin's Soviet heirs were indeed prepared to be more flexible than he in permitting considerable domestic autonomy and domestic setting of priorities in the people's democracies, they, too, would not tolerate any dissolution of Communist rule or any unilateral flight out of their hegemonic orbit. The Soviet empire in East Central Europe might change, but it would remain an empire in essential ways.

4

The next people's democracy to test these parameters and perimeters of post-Stalinist Soviet flexibility in and over East Central Europe was Romania. We left that country in the early 1950s, with its polity totally controlled by the Communists, its economy locked into the systemic machinery of mature Stalinism, and its internal power struggle won by the relatively nativist faction of Gheorghe Gheorghiu-Dej (see Chapter 3, section 6; Chapter 4, section 3).

Whereas in Poland, the post-Stalinist decompression was facilitated by Bierut's death, and in Hungary, it was sabotaged by the unregenerate Rákosi, in Romania it was orchestrated by Gheorghiu-Dej himself, who, at a plenary session of his party's Central Committee, on August 19 and 20, 1953, took the lead in political self-criticism, in initiating collective leadership, and in acknowledging excessive past fixation on heavy industry to the detriment of agriculture and consumption. Forced-labor camps were closed, and their biggest "employer," the Danube–Black Sea canal project—a cloaca of immense human suffering and mortality—was suspended in 1954. Stalin's heirs in Moscow now facilitated Gheorghiu-Dej's efforts at reform and at deepening his regime's precarious legitimacy by agreeing to dissolve the much resented, exploitative, joint Soviet–Romanian corporations, which since 1945 had enabled the Soviets to administer directly important sectors of the Romanian economy.

Whereas Gheorghiu-Dej had initially heeded Moscow's post-Stalin requirement to separate state and party offices by retaining the governmental premiership while yielding the party first secretaryship, he reversed himself on October 5, 1955, in preparation for the long-overdue party congress that met from December 23 to 28, 1955—the first since the coerced merger of the Social Democrats and the Communists in 1948. This congress was notable less for its predictable ratification of Gheorghiu-Dej's emulation of aspects of the contemporaneous Soviet "thaw" than for its packing of the Central Committee and Secretariat with his screened supporters and its elevation to the Politburo of three of his protégés, including his eventual hand-picked successor a decade later, Nicolae Ceauşescu.

The Romanian leaders navigated the perilous political seas of 1956 skillfully and craftily. They finessed Khrushchev's attack on Stalin with the claim that in their own country the problem of Stalinism had been solved in 1952 with the purge of the trio of Pauker, Luca, and Georgescu. They rationalized their collaboration with the Soviet suppression of the Hungarian Revolution by insinuating that this revolution incubated an aggressive Hungarian national irredentism, and therefore its success would have threatened Romanian security in Transylvania. Indeed, they subsequently exploited this alibi to whittle down the rather extensive Magyar cultural and administrative autonomy in Transylvania, simultaneously evoking and gratifying a quite truculent Romanian ethnonationalism that in the next decade was to be turned against the Soviet Union as well. Meanwhile, Moscow was sufficiently impressed by Romanian resoluteness and reliability to withdraw its military garrisons in the summer of 1958. Of course, given their long common border, Romania remained within striking distance of Soviet forces, if necessary.

The background to the Soviet–Romanian tensions of the 1960s lies in the grievances and aspirations generated by expectations of change within a context of relative backwardness. Stalin had insisted that each people's democracy press the rapid industrialization of its economy, particularly emphasizing its heavy metallurgical industry, and had discouraged multilateral cooperation and regional integration of the East Central European economies (see Chapter 4, section 1). Though distortive and wasteful, this Stalinist prescription had suited the Romanians well, for it implied that their relatively backward but resource-rich country would be enabled to realize its development potential and to cease being an undercapitalized, underproductive dependent of more advanced economies. Therefore, when Stalin's Soviet heirs eventually decided in June 1962 to revitalize the Council for Mutual Economic Assistance (CMEA), which Stalin had formed in January 1949 as his organizational retort to the Marshall Plan but had left dormant, and to hone it into an instrument for integrating the economies of the people's democracies through the "international socialist [that is, intrabloc] division of labor" (specialization in the members' respective marginal advantages), the Romanians understandably recoiled. They feared that this new program would effectively cheat them of their goal of comprehensive industrialization and modernization, and relegate them once again to their traditional and resented fate of being the backward supplier of raw materials and agricultural products to more advanced economies—this time those of the Soviet Union, Czechoslovakia, and East Germany. In flatly rejecting this fate and Moscow's apparent move toward it, the Communist regime and the Romanian people finally found themselves authentically on common ground. Moreover, to their emphatic satisfaction, they found themselves supported by Maoist China, whose leaders were similarly concerned about such incipient Soviet reneging on a commitment to help relatively backward Communist countries achieve rapid and all-around modernization.

Not only popular support and Chinese benevolence fortified Gheorghiu-Dej and his colleagues in their resistance to Moscow on an issue that they deemed vital. Romania's substantial energy resources and the availability of Western capital goods and Western trade options also helped. From the early 1960s, Romania turned to Western Europe and the United States for the machinery and equipment required by its unabating industrialization drive, and it found these goods to be better, cheaper, and delivered more promptly than Soviet bloc items.[9] And new cohorts of managers, technocrats, skilled workers, specialists, and professionals soon mastered and harnessed this Western technology to propel the economy into a high state of buoyancy throughout the decade. During the 1960s and most of the 1970s, Romania boasted the fastest industrial growth rate in the entire Soviet bloc. All this, in turn, enhanced the confidence of the political leadership in risking Soviet and CMEA disapproval in an era of proclaimed Soviet–American coexistence and of palpable Sino-Soviet friction that increased Romania's leverage and room to maneuver.

This partial disengagement from the hitherto well-nigh smothering Soviet economic embrace was complemented in the diplomatic and cultural arenas. Whereas in the 1960s the other people's democracies (except Albania) supported the Soviet Union in its open quarrel with Communist China, Romania remained neutral. Romania also reduced its participation in joint WTO military maneuvers and declined to let any be held on its own soil. In November 1963, Gheorghiu-Dej paid a state (not a party) visit to Belgrade, where Tito—who had recently again been at odds with Moscow—gave him the special honor of inviting him to address the Yugoslav National Assembly. They also agreed to cooperate outside the CMEA framework to develop the navigational and hydroelectric potential of the Iron Gates gorge in the Danube River. (But both then and later, Romania shunned Titoism's experiments with structural decentralization and pluralism.) Earlier in 1963, Romania had ceased jamming Western radio broadcasts; had opened itself to Western tourists, lecturers, exhibitions, publications, films, and dramas; had released hitherto banned "ideologically unacceptable" works by Romanians, such as the prewar nationalistic histories of Nicolae Iorga and the avant-garde dramas of the émigré Eugène Ionesco; had converted the Bucharest Institute for Russian Studies into one of four divisions of the new Institute for Foreign Languages; had similarly replaced the capital's Russian bookstore with the Universal Bookstore; had downgraded Russian-language study from a compulsory to an elective school subject; and had renamed many public streets, buildings, and institutions from Russian to Romanian eponyms.

Thus in several dimensions—economic, political, diplomatic, and cultural—the Romanian Communist leaders prepared new bases of legitimation and extended new bridges to their national public. But in contrast to the Hungarian Revolution, this Romanian desatellization was undertaken without any dissipation of exclusive Communist party sway over the society; without the cadres being swamped by the masses; without popular spontaneity breaching the apparat's controls; without traumatic rifts and changes within a tough, tight leadership; without public humiliation of the Soviet Union, denunciation of the WTO, or proclamation of neutrality; without spawning an ideological heresy; and thus without furnishing Moscow with sufficient provocation to invade Romania and abort this quite deft, relatively unassuming, even somewhat ambiguous maneuver to achieve national autonomy within, rather than against, the Soviet bloc and the Communist world. Maintaining this autonomy in later years has required constant skill at testing Soviet limits and foiling attempted Soviet encroachments.

For all its impressive industrial growth, the Romanian economy continued to be plagued by its traditional Achilles heel—low agricultural productivity. The "achievement" of full collectivization by the spring of 1962, though a political, administrative, and ideological success, only compounded the economic problem, for collective farms are notoriously less efficient and productive than private ones. Though quantitatively, Romania remained the Soviet bloc's second agricultural producer (after the Soviet Union itself), qualitatively it registered the region's lowest productivity per cultivated area and lowest level of agricultural mechanization.[10] In Romania, as elsewhere, collectivized agriculture has been Soviet-style Communism's biggest economic failure, and by the late 1970s, this failure would also contaminate and impair the earlier spectacular but one-sided industrial triumphs. Indeed, with the malaise in agriculture compounded by flawed management of industrial investment, staggering foreign debt, and a severe energy crisis, the Romanian economy descended into grave difficulties in the 1980s, and the resultant chronic austerity debilitated the regime's popular support.

But though there have been occasional strikes in Romania over bread-and-butter "workers' issues," they have not catalyzed the kind of worker–intelligentsia alliance pressing for socioeconomic as well as political reforms that was forged in Poland between 1976 and 1980 (see Chapter 6, section 1). The Romanian intelligentsia lacks the traditional aura of being deemed the conscience of the nation that is borne by the

Polish and Hungarian intelligentsias, and it has remained rather aloof from the grievances of the peasants and workers. And the churches (again in contrast to Poland) are weak and disorganized.

As expected, Nicolae Ceauşescu succeeded Gheorghiu-Dej as party leader after the latter's death from lung cancer on March 19, 1965, and he became head of state in December 1967. Ceauşescu soon imposed his own style of personalism, romanticism, and nationalism on the country. In the summer of 1965, he changed the party's formal name from "Workers" (adopted when the Social Democratic rump was absorbed in 1948) back to the more revolutionary-sounding "Communist" and upgraded the state's ideological-ceremonial appellation from "People's Republic" to "Socialist Republic," simultaneously conferring on it a corresponding new constitution. Though the beneficiary of his late mentor's patronage, Ceauşescu distanced himself from Gheorghiu-Dej's nimbus by rehabilitating several victims of his purges, criticizing him for arbitrary abuses, and initially presenting himself as more liberal and rational. But once his own power was consolidated, Ceauşescu outdid his predecessor in personal authoritarianism and compounded it with unprecedented nepotism. He not only combined a plethora of offices in his own person, but also promoted his wife, a son, several brothers, in-laws, and cousins to positions of power so blatantly as to undermine the institutional apparatus of the party and the state and to prompt the coining of the tongue-in-cheek label "dynastic socialism" to designate this process, for the Ceauşescu clan seemed indeed to be aiming at hereditary power.[11]

This process entailed ideological consequences. From the mid-1970s onward, the Ceauşescu regime sought to legitimate itself less in terms of the Marxist-Leninist concepts of leading class and vanguard party than with rhetoric celebrating Ceauşescu's supposed embodiment of the historical, eternal, and progressive virtues of the Romanian nation. Granted that all Communist regimes tend to court nationalistic sentiments and to cultivate ethnonational heritages at a certain stage in their postrevolutionary development, there is no denying that in Romania the Ceauşescus imparted to this stage an idiosyncratic twist that was both ludicrous and sinister. But for all their excrescences, distortions, repressions, and blunders, they remained committed to the goal of protecting Romanian autonomy and sovereignty vis-à-vis the Soviet Union within the Soviet orbit. Whether or not this stance will continue to secure their domestic legitimacy depends on the extent to which appeals to nationalism will become subject to the law of diminishing returns in a context of aggravating economic deprivation and police repression.

5

Czechoslovakia followed Romania in seeking to carve out more political and sociocultural elbowroom without provoking Soviet punishment. Though its Communist leaders abided by the presumed rules of the game that had supposedly emerged from the Polish and Hungarian crises of October and November 1956, by shunning any dissolution of domestic Communist rule or unilateral leap out of the Soviet orbit, their bid was traumatically rebuffed and suppressed with the Soviet invasion of their country (accompanied by East German, Polish, Hungarian, and Bulgarian units) in August 1968. Hence the entire episode of the "Czechoslovak spring" of 1968, at once edifying and tragic, cast a pall of uncertainty over the putative lessons of 1956.

Our earlier narrative had taken Czechoslovakia through the Communist party's seizure of power by the skillful application of nonviolent force in February 1948, followed by the imposition of a highly repressive regime that contrasted vividly with the seeming Communist moderation of the years 1945 to 1947, and culminating in an orgy of intraparty purges between 1949 and 1954 that left the country as, arguably, the most Stalinist of all the people's democracies, with Albania as its closest competitor for this dubious distinction (see Chapter 3, section 3; Chapter 4, section 3).

This dogmatic Czechoslovak Stalinism endured intact long after Stalin's death. Throughout the 1950s, the country remained sealed against any spillover from the contemporaneous Soviet "thaw" and the New Course decompressions that were agitating, lacerating, and/or exhilarating the other people's democracies. This inertia was due in part to the fact that all the surviving Czechoslovak Communist leaders were tainted by direct implication in the Stalinist terror and purges, toward which they nursed a custodial attitude. This party had neither a guilt-free senior counterpart to Gomułka or Nagy nor an incumbent leader with Tito's or Gheorghiu-Dej's penchant for autonomy. Thus, for example, in January 1957—after the Polish and Hungarian crises of the previous autumn—party chief Antonín Novotný could still denounce even the word *de-Stalinization* as being synonymous with "weakness and yielding to the forces of reaction," while as late as the end of 1961, he would dismiss as "irresponsible" petitions for a review of the ferocious purge trials of 1949 to 1954.[12] The only revision of this unyielding stance that Novotný and his cohort were prepared to concede was the rather grotesque one of deleting the count of Titoism from the charges against Rudolf Slánský and his co-defendants (out of deference to Khrushchev's current wooing of Tito) while declaring them justly condemned and executed on all other counts. Some surviving victims of the purges were quietly released from jail during 1956 and 1957, but without publicity, apology, exoneration, or rehabilitation. All this was, of course, in sharp contrast to the anguished soul searching that wracked the Polish and Hungarian Communist ranks during these years.

Novotný's obduracy was facilitated by the economic stability and growth that Czechoslovakia enjoyed during the 1950s. This not only served in general to smother political effervescence and dissent, but also specifically depoliticized the workers and thus forestalled the kind of turbulent alliance among workers, intelligentsia, and students that proved so portentous in Poland and Hungary during this decade. It also emboldened Novotný to declare that Czechoslovakia had successfully moved from the intermediate historical stage of people's democracy to the higher one of socialism—the first Soviet ally to do so—and to confer on the country a corresponding new, "socialist," centralistic constitution on July 11, 1960.

Only direct Soviet pressure finally budged the Czechoslovak leaders from their rigid dogmatism. Having renewed his attack on Stalinism at the Twenty-second Congress of the Soviet Communist party in October 1961, Khrushchev extracted from the reluctant Novotný a review of the trials of the 1949 to 1954 period. After more procrastination, the review board was finally appointed on August 30, 1962. Compounding the incendiary potential of this time bomb was the onset, also in 1962, of grave regression in the hitherto thriving, exemplary Czechoslovak economy—a development profoundly damaging to the regime's prestige. And now the Slovak Communists, who had long been subjected to relentless centralizing pressures and humiliation by their disdainful Czech comrades, decided to exploit the travails of their most overbearing tormentor, Novotný, to settle accounts. Their lever in this campaign to disgrace and dislodge Novotný were the Slovak and Czech critical writers and philosophers, who were given access to Slovak party publications in which they articulated ethnonational, humanistic, and ur-Marxist politico-cultural grievances. In response, Novotný sought to activate the latent anti-intellectualism of the workers—with some success.

Novotný's first serious organizational defeat was his inability to prevent the replacement of a close protégé by his critic Alexander Dubček as first secretary of the Slovak section of the Communist party in April 1963. Storming out of the meeting in a huff, he never again deigned to attend a plenum of the Slovak Central Committee. Later that same month came the Slovaks' preliminary but decisive exoneration and rehabilitation of their coethnic victims of the 1949 to 1954 purges, deliberately done in advance of Prague's decision on the review board's recommendations. A year later, in May 1964, the Slovaks extracted a formal restoration of some of their autonomous public institutions that had been closed by Novotný's *soi-disant* socialist constitution of 1960 (but as yet no clear substantive restoration of these institutions' real powers, which had been incrementally whittled away ever since 1946). By the end of 1966, the Slovak section of the Communist party had removed from its Presidium and Secretariat all the Prague-oriented, centralizing, terror-implicated satraps whom Gottwald and Novotný had imposed on it over the previous two decades.

Contemporaneous with this Slovak assertiveness, and also discomfiting to Novotný, was pressure by the legal profession throughout the country to restore the rule of law in its classic sense, with protection of the rights of citizens and defendants, with independence of courts and judges, and without the Stalinist presumptions of guilt by probability, analogy, or class background. This pressure yielded a series of laws and decrees between 1963 and 1966 that went far toward correcting the "distortions of socialist legality" that had run rampant since 1948. Of course, though very important, these juridical rectifications were not a substitute for political, administrative, and socioeconomic reforms.

Though reformist pressure also extracted some changes in the economic system, entailing less comprehensive centralized planning, fewer and less rigid control indicators for plant managers, more flexible prices, and realistic incentives—palpable, albeit still hesitant, acceptance of some market mechanisms—the reluctant Novotný made no secret of his hopes that these reforms would fail. And here he found it even easier to tap the anxieties and resentments of the workers than on the anti-intellectual front, for many industrial workers felt personally, materially, and immediately threatened by the reforms' partly intentional, partly unintentional consequences: wage differentiation, quality-work expectations, job insecurity, plant closings, even unemployment in inefficient sectors, price rises, and so forth—which scarcely corresponded to their customary sense of "socialism."

By the winter of 1967/68, most of the Czechoslovak reforms had come to a turning point where they would have to be either pressed more vigorously if their salutary potential was to be achieved or rolled back to halt the disarray that they were generating. Having managed to blunt several of the reforms but not to stifle the movement in toto, Novotný sought to capitalize on the workers' unease by attempting a coup late in December 1967. Some armored units were readied, but the army's Main Political Directorate foiled their use. The security services were more under Moscow's control than Prague's, and the Kremlin, though well disposed toward Novotný, did not want to become so blatantly embroiled in the personnel disputes of a traditionally reliable Communist party. The attempted coup was thus thwarted.

Novotńy was thereupon replaced as party first secretary on January 5, 1968, by the Slovak Alexander Dubček. Then forty-six years old, having spent thirteen years of his youth (1925–38) in the Soviet Union and another three (1955–58) as an adult student at the Soviet Higher Party School in Moscow (which graduated him with honors), a believer in the Czechoslovak reforms yet a veteran of the party apparat, a friend of the liberal intellectuals but not a radical critic of the system, a Slovak patriot now promoted to central responsibility, free of taint in the terror of 1949 to 1954, Dubček seemed an eminently prudent and promising choice, quite reassuring to the Soviet leaders. His selection was supplemented on March 30 with the elevation of the retired General Ludvík Svoboda to the presidency of the republic to replace Novotný, who had been pressured into resigning that office on March 22. Having fought on the Russian side in both world wars and having then facilitated—as the nominally nonpartisan defense minister—the Communist seizure of power in February 1948, Svoboda, too, was intended to reassure Moscow. And, indeed, the Soviets initially accepted the new Czechoslovak team quite cordially.

Dubček staked himself to three balanced and difficult commitments: (1) renovate the country's polity and revitalize its economy through the initiative of the Communist party; (2) earn public confidence in the reality of these reforms by systematically institutionalizing them, including the operational emancipation of state and social institutions from the strait jacket of stifling party controls; and (3) contain the resulting effervescence from spilling over the post-1956 Soviet perimeters of permissibility.

Always professing his commitment to Marxism-Leninism, to the leading role of the party (a role that had, however, to be earned), and to Czechoslovakia's intrabloc obligations, Dubček nevertheless authorized a vast public debate in the spring of 1968 that soon snowballed into a paroxysm of national revulsion against the two preceding "black decades." A stream of dismissals and resignations (as well as several suicides) ensued in the ranks of Novotný's old guard as one after another public institution and governmental agency, including even the Interior and Defense ministries, were taken over by the party's reformers. A presidential amnesty of May 9 and a rehabilitation law of June 25 finally restored liberty and honor to the living and dead victims of the terror and the purges. Censorship was tacitly ended in early March and formally abolished in late June. Officials mixed with the masses and submitted to authentic, probing criticism. Responsible interest groups and elective organs flourished as the "transmission-belt" tradition of the Stalinist years crumbled. But none of this exuberant behavior ever slipped the bounds of institution-focused legality, nor was it formally or substantively directed against "socialism" or the Communist party as such, nor was its tone anti-Russian, nor did it advocate restoration of capitalism or decollectivization of agriculture, nor was there any serious pressure to restore real autonomy or leverage to other, by now bogus parties—a notion that Dubček categorically rejected. Hence despite worried warnings from their Polish and East German clients (whose rigid, aging, factious, and unpopular elites understandably feared contagion), the Soviet leaders for long remained remarkably patient with the Czechoslovak experiment in "socialism with a human face," initially apparently viewing it as a flawed but tolerable effort at self-renewal.

Not until the end of April did Moscow express its first reservations, in the form of concern lest the well-intentioned Dubček become the captive of "antiparty elements." A series of intrabloc conferences and bilateral visits ensued during which the Soviet stance toward Czechoslovakia oscillated between pressure and understanding. This hesitancy and convolutedness of Soviet attitudes may have deluded Dubček into misjudging the extent of his freedom of maneuver. But the very fact that it did entail intermittent and escalating pressure (editorial and oratorical warnings, troop movements and maneuvers) aroused the Czechoslovak public into applying its own counterpressure on him to prove his patriotic and reformist credentials. Thus by the high summer of 1968, Dubček was trapped in a tightening vise. Though he might formally demonstrate that the Czechoslovak reforms were compatible with the canons of Marxism, hopefully claim that they were actually reviving the popularity of socialism, and modestly disavow that they were intended as a model for any other country, such logical and rhetorical adroitness achieved ever-diminishing returns from the Soviet leaders, who became increasingly uneasy about the spontaneous public energy behind the reforms and apprehensive lest their other East Central European dominos fall. And Czechoslovakia was, after all, a western and highly industrialized domino, the only one sharing borders with both Germanies and the Soviet Union itself.

The Soviet leaders appear to have felt their first sharp alarm over the "Two Thousand Words" statement issued by Czechoslovak intellectuals on June 27. A stinging indictment of two decades of party dictatorship and a ringing demand for accelerated reform, it juxtaposed endorsement of Dubček with a warning against yielding to external pressure. Though disavowed by Dubček as gratuitously provocative, it achieved much popularity and angered Moscow into delaying the withdrawal from Czechoslovakia of Soviet troops who had been engaged in WTO maneuvers there. On July 3, Soviet leader Leonid I. Brezhnev warned that "we cannot remain indifferent to the fate of socialism in another country," and by July 11, *Pravda* was ominously comparing the situation in Czechoslovakia with that in Hungary during October 1956 (though neither then nor later was there any violent turmoil in Czechoslovakia). On July 15, the WTO members (except Romania) collectively demanded that Czechoslovakia reimpose censorship, curb its intellectuals, and reverse its institutional liberalization. Prague's dignified and firm rebuttal of this ultimatum on July 18 prompted a surprising Soviet invitation for bilateral discussions.[13] Apparently, the Kremlin's collective leadership was having difficulties bracing itself to cross its Rubicon.

Nevertheless, the vise on Dubček was tightening relentlessly. On the one side, the Soviet Ukrainian leadership had now caught the acute Polish and East German alarm over contagion from the Czechoslovak fever. On the other side, his own reformist colleagues were warning Dubček that "the nation and the party will sit in judgment" were he to waver under pressure. Hence the seeming reprieve of the bilateral Soviet–Czechoslovak negotiations, which took place from July 29 to August 1, proved abortive, as neither side could yield. The last straw and the likely precipitant of the tortuous Soviet decision to invade Czechoslovakia and halt its dangerous experiments may have been Prague's publication on August 10 of a draft of new party statutes to be formalized at an Extraordinary Party Congress scheduled for September 9. They would have required the election of party officials by secret ballot and permitted minorities within the party to maintain and defend their political positions—devastating apostasy from Leninist orthodoxy.[14] The Soviets now hoped and expected that by administering a sharp shock they could prompt the residual conservatives in the Czechoslovak party's Central Committee (still almost half its members) to rise up against these draft statutes and the reforms in general, precipitate the "faithful" workers into helping "unmask the dirty intrigues" of the high-riding intelligentsia, and peel the more nationalistic Slovak Communists and intellectuals away from Dubček and from the general reformers through substantive federalization and autonomy. Armed with this complex of exasperations and calculations, the Soviets and their client-allies invaded Czechoslovakia during the night of August 20/21.

Though the invasion was met with massive nonviolent civil (but no military) resistance, which also encompassed workers who had hitherto been wary of the reform movement, and though Dubček formally remained party first secretary until his replacement by Gustáv Husák (also a Slovak) on April 17, 1969, the "Czechoslovak spring" was effectively blighted. Through the invasion, Moscow halted a process that it had come to perceive as ideologically dangerous and politically contagious; reasserted its global power credibility and regional hegemony; braced its Polish and East German clients, intimidated its Hungarian one, deflated the Yugoslavs, sobered the Romanians, and shocked the Albanians; and punctured the confidence of American diplomatic "bridge builders," yet without extinguishing their ongoing interest in détente.[15] Compared with these gains, the Kremlin deemed the political costs of the invasion to be acceptable and shrewdly anticipated that many of them would evanesce over time. And, indeed, from a long register of seemingly prohibitive costs at the time, a few real ones remain two decades later: (1) A Czechoslovak consensus, dating from the Munich calamity of thirty years earlier, that national independence and dignity are linked to alliance with the Soviet Union was shattered, and the traditionally Russophile population of a strategically and economically important Soviet client was estranged. (2) Lingering post-Stalinist hopes throughout East Central Europe that the Soviet bloc might yet be transformed into a genuine socialist community and partnership were shattered; instead, the "Brezhnev Doctrine" rationalizing the invasion confirmed that Moscow deems its client-neighbors to be less sovereign than non-Communist states. Yet neither the invasion nor the doctrine could ensure the lasting political stabilization of the area.[16] (3) The

Western European Communist parties, which had been striving mightily to project a more democratic and responsive image in their countries, were embarrassed, offended, and weakened. (4) The ossified Soviet political elite's allergy to modernism, liberalism, democratization, Marxist humanism, and non-Soviet nationalism was again exposed, and the Russian people's xenophobic revulsion for the "ungrateful and unreliable" client-allies was unhealthily reinforced. (5) Apart from catalyzing the Czech–Slovak structural federalization of the state and government (legislated on October 27, 1968, and implemented on January 1, 1969), the Soviet invasion solved none of the other festering socioeconomic and politico-cultural problems that had prompted the reform movement; rather, by inducing cynicism and apathy, it aggravated them. Indeed, even the federalization was not as politically constructive as it might have been, for the circumstances under which it was finalized left the Czechs with a lingering resentment that the Slovaks had purchased it at the cost of general democratization—a sentiment that now reciprocated the Slovaks' earlier feelings of betrayal at the hands of the Czech centralizers during the Stalinist years as well as during the interwar decades.

It remains to draw a general lesson from the Soviet suppression of the "Czechoslovak spring" of 1968 comparable with the lessons of 1956 from Poland and Hungary. The lesson is that, pending profound changes in the Soviet Union itself, a liberalizing "renewal" of sociopolitical life in a neighboring Communist state, even when initiated by a fellow Communist party expressly committed to the WTO, to CMEA, to bloc solidarity, and to its own domestic monopoly of power, is, in the final analysis, still unacceptable to the Soviet leadership because such a process cocoons the specter of a possible loss of control to social spontaneity and organizational autonomy. This lesson should be sobering, but not surprising.

6

Albania had actually preceded Romania and Czechoslovakia in confronting the Soviet Union, but as its leaders had been motivated by nationalistic and ideological anxiety about Soviet betrayal rather than by any urge to test the post-1956 perimeters of Soviet political permissiveness, the following account of Albanian assertiveness has been postponed to here.

The linked Soviet–Yugoslav and Albanian–Yugoslav rifts in 1948 had been followed by the conventional internal purge of the Albanian Communist party (renamed the Albanian Party of Labor in September 1948) and by the country's translation from a satellite of Yugoslavia to a satellite of the Soviet Union (see Chapter 3, section 8). The constitution of March 1946, modeled after that of Yugoslavia, was amended in July 1950 to approximate the Soviet one more closely. Albania was admitted to CMEA in February 1949 and became a founding member of the WTO in May 1955. During these years, its leader, Enver Hoxha, also espoused the basic political, socioeconomic, and cultural-ideological institutions and programs of Stalinism. Unlike most of the other peoples of East Central Europe, the Albanians assessed this Soviet embrace as protective (vis-à-vis Yugoslavia and Greece) rather than as smothering, as enhancing their modernization rather than as exploitative. Therefore, Khrushchev's commitments in the mid-1950s to reconciliation with Tito and to denunciation of Stalin were profoundly disturbing to them and especially to their rulers. In wooing Tito, the Soviets appeared willing to sacrifice Albania's security and independence to this presumptively predatory neighbor, and in repudiating Stalin, to abandon the values and policies that were propelling Albania's modernization. And Khrushchev's readiness in May 1960 to discuss the possibility of autonomy for the Greek minority in southern Albania with a Greek political leader (Sophocles Venizelos of the opposition Liberals) inevitably compounded Albanian anxieties, as did his curtailment of Soviet grain shipments, withdrawal of technical specialists, suspension of economic credits, and other minatory decisions and gestures.

As the enraged but unintimdated Albanian Communist leaders responded with sharp ideological reproaches, Khrushchev first attempted a covert coup against them in July 1960, using the Albanian navy and a pair of disgruntled senior party officials. After this failed, he switched to a public attack at the Twenty-second Congress of the Soviet Communist party in October 1961, execrating Hoxha and his cohort as slandering Judases more malicious than even open enemies and suggesting that the Albanian people overthrow them. Khrushchev's diatribe was reminiscent of Stalin's equally vain appeal of 1948 to the Yugoslav party to replace Tito and his entourage (see Chapter 4, section 2). And in December 1961, Khrushchev went even further than Stalin had done by severing state relations with the target of his scorn.

In the context of the aggravating Sino-Soviet rift, Maoist China promptly compensated Albania for this Soviet ostracism by rallying to it with extensive ideological, political, technological, and economic support. And Hoxha would soon reciprocate by emulating important aspects of Mao's "cultural revolution" in order to "cleanse" Albanian society of unsocialist tendencies, prune the bureaucracy, and guard the party against revisionist degeneration. But this Albanian version of the "cultural revolution," unlike the Chinese one, neither slipped out of the party's control nor developed any spontaneous momentum, and it was not marked by anomic violence, public humiliations, or economic disruptions.[17] Throughout most of the 1960s, these two allies in the Communist world, so unequal in power, were agreed on that world's basic judgmental issues: the nature of imperialism, the alleged degeneration of the Soviet leadership, the supposed fraud of peaceful coexistence, the sanctity of class struggle, revolutionary militancy and proletarian dictatorship, the criteria for assessing authentic communism, and the need to unmask revisionist deviations.

Under the umbrella of this ideologically powerful yet geographically remote Chinese patronage, the Hoxha regime came to perceive the Soviet Union rather than Yugoslavia and Greece as the main threat to itself and its country. This perception was reinforced from several directions in the mid-1960s. First, after the fall of Yugoslavia's secret-police chief, Aleksandar Ranković, in July 1966 (see section 7), Belgrade eased its heavy hand toward Yugoslavia's Albanian minority and sought reconciliation with Tirana. Second, the Greek colonels' junta that seized power in Athens in April 1967 proved surprisingly less militant than its monarchist predecessors on the issue of Greek claims to southern Albania. Third, the invasion of Czechoslovakia in August 1968 by the Soviet Union and its other satellites intensified Albanian fears of aggression against itself from this same quarter. Hence to protect itself formally against a casuistical Soviet exploitation of the WTO similar to the one Brezhnev had used to rationalize his crushing of Dubček, Albania unilaterally withdrew from that organization on September 12, 1968, pithily charging that the recent Soviet invasion of Czechoslovakia had perverted it "from a defense pact against imperalist aggression into an aggressive pact against the socialist countries themselves."[18] To emphasize the seriousness of this decision, high-level Albanian and Chinese military-cum-Politburo delegations exchanged visits in October and November 1968.

The Hoxha regime now sought to prepare its people for a possible Soviet invasion from Bulgaria and/or the sea. Logically, albeit surprisingly, it also pledged to aid Yugoslavia and Romania in case they came under such a Soviet attack. Aggravating the Albanian war psychosis was a disappointed perception that China had shown itself to be weak in the face of Soviet belligerence during a series of Sino-Soviet border clashes in the spring of 1969. This, in turn, precipitated a major Albanian reassessment of the long-excoriated West, which now appeared to be the only remaining actor able and willing to deter Soviet aggression.

Albania's ensuing supplementation of its hitherto exclusive link with China with an "insurance" opening toward the West was neither an easy nor a steady process. Rather, it was prolonged, fitful, and marked by relapses and hesitations. Ideologically, the Albanian leadership remained highly sympathetic toward Mao's radicalism and fundamentalism, but pragmatically it realized that precisely these qualities, due to the turmoil they were inflicting on China, attenuated its ability to subsidize Albania economically. And China's geographical remoteness limited the direct military aid that Albania could expect from it in case of a Soviet attack. (From the Chinese perspective, the value of this small ally, which had been high as long as the Sino-Soviet rift remained intensely ideological and confined within the Communist camp—because it served as proof that China was not alone—slipped with the Chinese decision in 1971 to globalize and semisecularize the quarrel by seeking a rapprochement with the United States. Implicitly partnered by the American superpower, China henceforth deemed its alliance with Albania as less important than it had been in the 1960s, though still worth preserving.) As for Albania's incipient accommodation toward the West, its leaders found approaching Europe to be ideologically less embarrassing than soliciting the United States.

Preparatory steps were the restoration of full ambassadorial exchanges with Yugoslavia and of diplomatic relations with Greece, respectively on February 5 and May 6, 1971. Thereupon, Albania's trade with Yugoslavia and non-Communist countries more than quadrupled between 1970 and 1975, and since the mid-1970s, about one-third of its foreign trade has been with the West. Hoxha also implicitly supported Yugoslavia (though he continued to criticze Titoism ideologically) in its perennial quarrel with Bulgaria over the Macedonian issue, as he viewed the latter state as a Soviet stalking-horse in the Balkans. Diplomatic contacts with many Western European states also intensified. Globally, the number of states with which Albania maintained diplomatic relations more than doubled between 1970 and 1980 from forty-two to ninety-four, and it developed commercial ties with about fifty. Meanwhile, the seeming ascendancy of the "Shanghai radicals," toward whom Hoxha harbored an emphatic ideological affinity, in China in the mid-1970s revived his hopes for the utility of this preferred China connection. Thus Albania, which had customarily been one of the international system's more isolated states, with but one patron at a time (prewar Italy, wartime Germany, then postwar Yugoslavia, the Soviet Union, and China) now enjoyed for an interval the unaccustomed luxury of having several irons in its diplomatic fire simultaneously.

The revival of the Sino-Albanian alliance was blighted by Hoxha's provocative reaction to the victory of China's pragmatists over the radicals after Mao's death in September 1976 and the resultant intensification of Sino-American cordiality. Though the Chinese, seeking to avoid an open rupture, initially turned the other cheek to the shrill and sustained Albanian denunciations of their new policies, their hand was eventually forced by Albania's gratuitous, emphatic endorsement of Vietnam in both its conflict of 1978 with China's protégé Kampuchea and its brief border war of early 1979 with China itself. Other than ideological hubris laced with a shrewd sense that the Soviet threat had abated, it is difficult to find a plausible explanation for Hoxha's conduct in this matter, for it finally stirred the long-patient but now exasperated Chinese to terminate all their economic, technical, and military assistance to Albania, simultaneously recalling their advisory specialists and expelling Albanian students from China. Formal, though henceforth frigid, diplomatic relations survived, but trade withered. Though the alliance had been ideologically and politically warm for most of its almost two decades, the absence of close cultural and emotional bonds between the two peoples and their respective elites now facilitated its termination without domestic repercussions in either country. China had never really penetrated Albania.

Despite this rupture of its alliance with China, Albania spurned repeated Soviet bids for reconciliation and made no overtures to the United States. Rather, it opted for a declarative stance of self-reliance while intensifying and diversifying economic relations with most European states (West and East) other than the Soviet Union, especially with Yugoslavia. Indeed, East Central Europe replaced China as a market for many Albanian exports, whose quality is too low to render them salable to the West. A certain degree of additional domestic belt-tightening was required due to the end of Chinese credits and subsidies; but as the Albanian people had not developed high consumption expectations, these costs and constraints were weathered without social disruptions or political reforms. That the rhetoric of self-reliance was not mere bombast is suggested by the fact that by the early 1980s, Albania had achieved self-sufficiency in the production of cereals, oil, electricity, and chemicals; was processing most of its minerals locally; and had developed a considerable industrial sector. Yet it remained Europe's least developed country, with a per capita gross national product of only $740 in 1978.[19]

Hoxha had never permitted Albania's external gyrations—from the Yugoslav, to the Soviet, to the Chinese unequal alliances, and then to his latest phase of self-reliance combined with multiple international contacts—to spill over into any domestic relaxation of his repressive Stalinist system. Thus the goals of the Albanian regime in seeking an end to satellite status and a recovery of national autonomy were quite different from those of Poland, Hungary, and Czechoslovakia and were more analogous to those of Romania. After smashing a rival pro-Yugoslav faction in 1948 and 1949 and nipping the attempted Soviet subversion by proxy in July 1960, Hoxha had felt domestically secure until the mid-1970s, when, in rapid order, he inflicted three more purges on his party and state apparatuses. The first, in the spring of 1973, was directed against a trio of Central Committee members whose responsibilities were in the fields of culture, education, propaganda, and youth activities and who appear to have recommended some loosening of reins in those fields in order to improve the morale of intellectuals, students, and young workers. The three were disgraced and relegated to manual labor. Next, in mid-1974, came the decapitation of Albania's military high command, whose craving for professionalism in lieu of "people's war" notions and whose corresponding wish to limit the authority of political commissars alarmed Hoxha as a mutinous effort to escape the party's control and to challenge its primacy. The veteran defense minister (who was also a Politburo member), the chief of staff, and the head of the army's Political Directorate were secretly executed. Finally, a year later, came the turn of the chief economic planners and administrators whose recommendations for greater decentralization in economic decision making and reassessment of targets in the light of an impending reduction of Chinese aid and credits (China being in the throes of severe retrenchment and having given due notice to Albania) similarly struck Hoxha as liable to destabilize his regime. They were eliminated between May and July 1975. For the next six years, Hoxha again felt secure—until 1981, when he turned against his longtime closest colleague and presumed political heir, Mehmet Shehu. Apparently fearing that Shehu was so unpopular that his succession might provoke turmoil, Hoxha decided to dislodge him as designated heir in favor of Ramiz Alia, whereupon Shehu committed suicide on December 18, 1981. The resultant purge of Shehu's family and followers took a year and encompassed the dismissal, arrest, and execution of a score of cabinet members, senior military and security officers, and party officials.[20]

Hoxha paralleled these periodic bleedings of the Albanian elite with an unrelenting hard line toward the masses, refusing them any political relaxation, any modification of agricultural collectivization, any easing of consumption, any slowing of the pace of industrialization, and any slackening of pressure against religion. He was convinced that this rigidity was politically functional as well as ideologically warranted. Where had popular de-Stalinization led the other people's democracies but into a deplorable morass of revisionism, revolution, counterrevolution, Soviet intervention, civil strife, and general miasma? Hoxha believed that by avoiding even the first, seemingly easy, step down this slippery path, he was ensuring his country the better lot of domestic stability and peace. Though others might demur that this was the peace of the graveyard, he deemed it preferable to ferment, for the concept of healthy or productive ferment had no place in his universe of political values.

Even when he might easily have purchased some authentic national popularity without sacrificing his absolute control, Hoxha hesitated to identify with ferment, even in a neighboring country. Thus he remained rather aloof from the rising ethnonational militancy of Yugoslavia's Albanians, who since the spring of 1981 have been locked in intense political combat, interspersed with occasional violence, with their Serb neighbors in the Kosovo province (Kosovë in Albanian) of Yugoslavia's Serbian republic, adjacent to Albania. Numbering somewhat less than 2 million in comparison with Albania's own population of just under 3 million, the ethnic Albanians in Yugoslavia are proportionately one of the world's largest irredentas. Though they are outbreeding and outsettling the Serbs, they regard themselves as politically oppressed, deprived, and intimidated as well as economically exploited by the Serbian republican authorities. Therefore, they seek the elevation of Kosovo to full and equal republican status, a demand that Belgrade fears it cannot concede without utterly alienating the already disgruntled Serbs, Yugoslavia's largest ethnonation. Both peoples regard Kosovo as their national cradle. Hoxha was quite restrained in extending even moral support to these freshly assertive Albanian coethnics across the border and scrupulously refrained from political intercession on their behalf. He made it clear that not only did he harbor no territorial aspirations for the Albanian state in this matter, but that any destabilization of the Yugoslav federation was unwelcome and could only serve sinister Soviet interests. Nevertheless, this fastidious forbearance did not spare Albania from being blamed by the exasperated Yugoslav authorities for their troubles in Kosovo.[21]

Enver Hoxha, born in 1908, died on April 11, 1985, after forty-one years in power—longer than any other Communist ruler except North Korea's Kim Il Sung. As prearranged, he was succeeded by Ramiz Alia, born in 1925 of parents who had fled from Kosovo into Albania after the Balkan Wars of 1912 and 1913. While Alia has expressed more intense interest in the fate of Yugoslavia's Albanians than had his mentor, he has yet to change the substance of Hoxha's policies, processes, and structures. His immediate priority is to project an impression of continuity and stability. Formally, juridically, and potentially, his power is formidable, for the constitution of December 16, 1976, which proclaimed Albania a socialist state (in the pattern of Czechoslovakia and Romania) also reserves to the Party of Labor sole political authority in the country and (uniquely) identifies the party's first secretary as commander in chief of the armed forces and head of the Defense Council.

7

In Yugoslavia, an ironic transformation took place between the 1950s and the 1970s. Having been the first client-ally to defy Moscow and to experiment with a different style of Communism and having then functioned as the inspirer, catalyst, and whetstone for a series of confrontations between several other people's democracies and the Soviet Union, Yugoslavia's own politics turned parochial in the process, as it ceased to be the Communist world's only boat rocker.

The Soviet–Yugoslav reconciliation of 1955 had been severely strained by the Hungarian Revolution of 1956, which each blamed the other for having provoked, mishandled, and aggravated. The media polemics between them grew so sharp that Khrushchev and Tito, each of whom had a personal interest in averting another total rift, met in August 1957 at Bucharest to patch up this quarrel—with mixed results. Yugoslavia now became the first non–Soviet bloc country to recognize the German Democratic Republic (East Germany), and Tito took to lauding proletarian internationalism and to vehemently criticizing NATO. Perhaps as a gesture to Moscow, perhaps only coincidentally, Milovan Djilas, who was already in prison for having utilized Western journals to criticize the Soviet crushing of Hungary, now had his jail term extended to nine years for having published *The New Class* in the West. But the Yugoslavs refused to sign a Soviet-drafted declaration issued by all the other ruling Communist parties to mark the fortieth anniversary of the Bolshevik Revolution in November 1957 that alluded to "the leading role" of the Soviet Union and condemned "revisionism in the guise of national peculiarities."

Khrushchev, who appears to have miscalculated that Tito's new readiness to distance himself from the West and to tighten domestic discipline implied a corresponding willingness to reenter the bloc on something like Gomułka's terms, was outraged by the Yugoslav rejection of the anniversary declaration. He soon denounced the Yugoslavs as a Trojan horse within Marxism and justified Stalin's original charges against them. The execution of Imre Nagy on June 16, 1958, at the height of these recriminations, was also a Soviet slap at Tito. Tito, too, had apparently miscalculated that his brief second flirtation with Khrushchev in the summer of 1957 would decisively facilitate the latter's mop up of residual Stalinists in the Soviet Union and catalyze a pivotal decentralization of the international Communist movement, leaving Yugoslavia as the holder of the balance. The upshot of these reciprocal miscalculations was a new round of *tu quoque* polemics and frosty relations that lasted for three years.[22]

An added complication was that the Soviets were being constrained by Maoist Chinese pressure on their left flank, as it were, to be more centralistic and less patient with the Yugoslav revisionists than they had been before the Hungarian debacle. Mao was insisting on ideological uniformity, strict bloc discipline, and Soviet leadership not because he had great confidence in Khrushchev's political wisdom, but because he sought a veto over Soviet quests for détente with the West (with which China's relations were then at a nadir); this he could have only if Moscow were accountable to a tightly organized bloc. In other words, a firmly centralized, orthodox bloc—albeit under Soviet leadership—would become Mao's mechanism for enchaining Khrushchev, and this, in turn, required a harsh line toward the wayward Tito. While Khrushchev, too, had outgrown his former regard for Tito as potentially helpful, he would have preferred to treat him as a minor gadfly rather than a major menace, as Mao insisted. Thus when the

Sino-Soviet rift unfolded during 1961 and 1962, the Soviets utilized their release from this Chinese embrace to arrange another reconciliation with Yugoslavia.[23]

But as this third reconciliation took place in the context of escalating Sino-Soviet enmity and of high Soviet–American tension (for example, the Cuban missile crisis of October 1962), followed by an erosion of Khrushchev's authority even before his ouster on October 14, 1964, the Yugoslavs took care to keep it in lower key than the first one of 1955 and 1956 and not to burn any other bridges in the process. Thus while on the one hand, Soviet–Yugoslav trade and mutual diplomatic support flourished during the mid-1960s (for example, during the Middle East crisis and Six-Day War of May and June 1967), on the other hand, Belgrade was simultaneously accelerating Yugoslavia's de facto economic integration into Western Europe (whither 1 million of its workers emigrated and whence they repatriated heavy remittances), licensing a general atmosphere of *embourgeoisement*, engaging in further market-oriented structural reforms of the economy, and adopting a quite pluralistic constitution on April 7, 1963—all flying in the face of Soviet ideological orthodoxy. This Yugoslav insistence on keeping the third reconciliation with Moscow cool yet authentic seemed vindicated by the Soviet invasion of Czechoslovakia in August 1968, which buffeted it even more severely than the invasion of Hungary twelve years earlier had strained the first reconciliation.[24]

After 1968, Soviet–Yugoslav relations eventually settled to a level of relative normality, with occasional ups and downs but without the extreme oscillations of the two preceding decades. The Soviet leaders at last learned to perceive Yugoslavia realistically as a middle-level, vulnerable, but politically important Balkan state. They no longer exaggerated its potential either as a threat to (Stalin) or as a crutch for (Khrushchev) the Soviet system. The Yugoslavs, in turn, came to treat the Soviet Union as but one of several pillars of their foreign policy, along with Europe, the Third World, the United States, and China—with which they finally achieved a rapprochement in 1971, after China's decision to globalize and pragmatize its foreign policy.[25] While Tito's leadership of the nonaligned camp continued to have an anti-American edge, he never again embraced the Soviet Union as a comrade. Nor did he any longer entertain his hopes of the Khrushchev era that its political system might qualitatively change. Meanwhile, he and his country also ceased to be an inspiration for dissidents and reformers in the Soviet bloc, where developments overtook the Titoist model.

The earlier sequential Soviet–Yugoslav rifts and reconciliations had influenced Yugoslav domestic politics in a complex pattern. At the risk of some oversimplification and telescoping, the dominant effect may be summarized as follows: each rift animated those local forces pressing for further structural liberalization, while each reconciliation played into the hands of those who wished to limit Yugoslavia's variance from post-Stalinist Soviet institutions and processes. Yet each of these major tides was also "dialectically" accompanied by a riptide. For example, immediately after the Hungarian Revolution, during a trough in Soviet–Yugoslav relations, Kardelj was arguing that further economic liberalization and political democratization were needed to preclude similar turbulence in Yugoslavia, while Ranković was simultaneously arresting and harassing critics lest the Hungarian turbulence prove contagious. Again, the onset of the third rift in 1958 was followed by a bout of intensified police interrogations and harassments of liberals and dissidents, to the dismay of many party (League) leaders.

Grave economic difficulties in the early 1960s alerted the Yugoslav leaders to the disquieting lesson that their vaunted "Titoist" reforms of 1950 to 1953—workers' councils, decollectivization, decentralization, social self-management, and so forth—had been hybrid comprises and were now threatening to fall between two stools. Though ideologically bold and politically innovative, they had left the economy and the state apparatus suspended between crippled central-planning institutions and distorted market mechanisms. The consequence was inefficient fragmentation of industry, uneconomic duplication of prestige projects, wasteful sprouting of "political" factories that could not be closed even though blatantly unproductive because they were jealously protected by republican and local governments, deep regional divergences in income, productivity, social services, and development, as well as sporadic strikes and widespread disgruntlement. Even more ominously, as the debate about how to correct these flaws proliferated downward and outward from the political leadership to the professional economists, lawyers, and political scientists, and then to the public at large, it exposed once again the raw nerves and wounds of Yugoslavia's interregional and interethnic tensions and hatreds, which had caused so much bloodshed during World War II and whose supposed overcoming was a central component in the Communist Partisan movement's "myth of political founding."

Three analytically distinct but politically intertwined issues thus came cumulatively to a head in this great debate: (1) whether to maintain, restrict, or extend the degree of personal freedom and political openness that had been achieved by the mid-1950s; the personal vicissitudes of Djilas and a few other intermittently jailed and released dissidents served as the bellwether of this issue; (2) where to stabilize the dynamic tension between centralization and federalization in institutionalized decision making, especially in matters of investment allocations; in an oversimplified but understandable fashion, the issue came to be likened to the prewar polarization between the Serb penchant for hegemony and the anti-Serb, Croat-led wish for autonomy; and (3) whether socioeconomic development would be best served by capitalizing on the advantages of the country's relatively advanced northern (ex-Habsburg) half or by promoting the underdeveloped southern (ex-Ottoman) half; this was often termed the efficiency-versus-equality dilemma. That the gap between the two halves should not widen further and should ultimately be closed for the sake of the federation's integrity and even viability was agreed in the 1960s, but selecting a rational and equitable strategy to achieve this goal proved profoundly conflictive, replete with ethnically loaded charges of exploitation, neglect, parasitism, colonialism, shackling, and extortion; in Yugoslavia, interregional discrepancies are inevitably experienced as interethnic tensions.

One reason why this compounded debate was so extended and proliferated is that Tito himself was for long equivocal in his assessment of all three dilemmas. His reticence is understandable, given the gravity of the stakes: how to achieve rapid modernization without social collapse; how to balance collective cohesion with individual liberty; how to reconcile a once revolutionay and still monopolistic ruling cadre to popular participation in a context of acute interethnic tensions and of profound interreginal disparities that conceivably might again disintegrate the state; how to steer between the Scylla of excessive control that was inducing stagnation and the Charybdis of political devolution that risked chaos.

By the mid-1960s, the centralistic, Serb-led, "southern," planned-economy coalition had fractured and hence lost this contest, largely because its Macedonian and Kosovo Albanian contingents had become persuaded that their underdeveloped constituencies were not really well served by a Serb-dominated, etatist, redistributing central apparatus that favored Montenegro, Bosnia-Hercegovina, and, of course, Serbia proper. Though the competing coalition—pluralistic, "northern," market-prone, "liberal"—was also differentiated between those who wished to devolve authority to the republics (the Croat and Slovene ethnonationalists) and those who wished to devolve it all the way down to the communes and enterprises (the marketeers), this division was at that time politically less consequential.

The fulcrum and symbol of the reformers' victory was the fall in July 1966 of their archfoe, the veteran Serb centralizer, security-police chief, and apparat-patronage controller Aleksandar Ranković. Predisposed by his entire career to favor a style of economic development that emphasized prestigious, large primary-industrial and infrastructural investment projects, Ranković was simply out of his depth in the reformers' new discourse about cost–benefit analysis and qualitative productivity indicators. Tito had long valued him for his loyal comradeship, Partisan courage, shrewd competence, disciplined toughness, in a sense even for the primitiveness of his vision of Communism. But when Ranković was shown to be manipulating the Serbian party cadres, state bureaucracy, and police network in order to paralyze federally mandated reform experiments, and even to be wiretapping his fellow leaders, Tito dropped him and his lieutenants. It was a mighty fall, for Ranković was then vice president of the Yugoslav state, one of a trio of secretaries to the Central Committee of the League of Communists (the party), and generally regarded as Tito's heir apparent. As with the repudiation of Djilas twelve years earlier, the personal intervention of Tito was needed to end an impasse, and, in doing so, he again demonstrated his political mastery of the party and the country.

The fall of Ranković not only resolved the great compounded debate in terms of policy, but also entailed institutional consequences. In its aftermath, the rogue secret-police apparatus was tamed, Serb preponderance was clipped, and hitherto intimidated ethnic groups such as Yugoslavia's Albanians were unbound. Yet in the midst of the resultant reformist euphoria, Tito prophetically gave notice that he had "not joined up with any liberals"[26]—an implicit warning not sufficiently heeded at the time.

Indeed, the victorious coalition of "northern," pluralistic, market-celebrating reformers now proceeded to overplay their hand as egregiously as had the recent losers. First, they had not really thought through the question of how to mobilize investment capital during a crucial transitional interval before market forces would "bite." Nor had they done their sums on the related problems of unemployment and the underdeveloped regions. Instead, they hastily enacted a large package of multiple reforms full of lacunae and contradictions that led to multiple irrationalities and by the end of the decade threatened to discredit the very concept of systemic reform. For example, in the face of tighter credit and looming unemployment, workers' councils and communes predictably took the shortsighted but easy route of restricting or even eliminating their allocations for research, development, and scholarships, dismissing high-salaried engineers and technicians, while retaining surplus workers. In the macroeconomic dimension, the second half of the 1960s was a seesaw period marked first by stagnation, unemployment, emigration, and declining real incomes, and then by inflation, galloping balance-of-payments deficits, and a weakened currency. During both phases, the politically dangerous regional disparities between the wealthier north and the poorer south widened. Second, there now surfaced the hitherto latent tension within the reformist coalition between those regional party barons who pressed to transfer centralized power to the 6 ethnonational republics and 2 provinces, and those "localists" who wanted to see it devolve all the way down to the country's 500-odd communes and their many self-managing corporate enterprises. In this competition, the first element was able to draw on more resonant and deeper emotive sources than its relatively cerebral and abstract competitor. In the process, it flirted with—and in Croatia allowed itself to be seduced by—some emphatically centrifugal, ethnonationalistic, explicitly anti-Serb, and implicitly anti-Yugoslav crypto-separatist nonparty forces. This prompted an inevitable blacklash when, in December 1971, Tito personally once more, after considerable procrastination, intervened to purge the runaway Croatian governmental and party apparatuses.

Yet Tito took care not to let the political pendulum swing back to overcentralization and police methods of governing or to a command economy and Serb hegemony. Tactically, he allayed several Croatian economic grievances, and strategically, he sought to point Yugoslavia toward a structural synthesis in which a revitalized, disciplined, trans-ethnic, and effective Communist party (League) would protect self-managerial autonomies and initiatives against centrifugal and centripetal ethnonational arrogations from any direction. Conflicting group pressures—an inevitable by-product of decentralization—would remain legitimate but be productively coordinated. To facilitate such coordination, there would be an "ethnic key" (quota system), ensuring a reasonably fair distribution of appointments among the country's constituent ethnonations. In effect, the regenerated party (League) would "hold the ring" within which technocrats and workers would manage the economy and society. Yet it would not be merely a referee and adviser in this process but would actively reengage itself in "cadre policy" in both governmental and economic institutions.

As with every previous systemic change, this new structural-political vision was now formalized in a new constitution, postwar Yugoslavia's fourth (not counting numerous intercalated amendments), promulgated on February 21, 1974. But this final effort by Tito to restore a healing balance fell short of his aim. It achieved institutional stabilization but failed to give the party (League) a real sense of historical direction. Nor did it cure the country's two recalcitrant and related malaises: interethnic stress and interregional inequality. And without such a cure, the vaunted "brotherhood and unity of the Yugoslav peoples" will remain precarious.

Indeed, even under the reactive constitution of 1974, the Yugoslav federation still manifested many features of a confederation. So extensive were the powers reserved to the six constituent republics and two autonomous provinces of Serbia that ambitious and talented politicians and administrators often resisted "promotion" to the center, deeming such transfers as tantamount to being severed from the real loci and sources of power. The republics and the provinces assigned highest priority to their own partial interests, and the limping national economy was managed and planned, if at all, only thanks to protracted interrepublican negotiations rather than through federal policy or local self-management.[27] No wonder that Tito, shortly before his death on May 4, 1980 (at age eighty-eight), regretted not having exploited his lancing of the Croatian abscess in 1971 more radically to subdue and leash the "eight little autarkies."

After Tito's death, the Muslim issue in Bosnia-Hercegovina, the new militancy of the Kosovo Albanians, and—in reaction thereto—an irate Serbian backlash replaced Croatian pugnacity as the most fevered expressions of Yugoslavia's chronic ethnonational malaise.

Between the late 1940s and the early 1980s, the Muslim proportion of the population of Bosnia-Hercegovina increased from roughly 30 to 40 percent, the rest consisting of Serbs and Croats in a ratio of approximately 2:1. These Serbocroatian-speaking Muslims were for long the country's least ethnonationally conscious segment—a condition encouraged until his fall in 1966 by Ranković, who viewed them as a religious, not an ethnic, group. But thereafter, they became more assertive, first in response to indigenous factors, such as the official Yugoslav legitimation of polyethnicity and their own educational and social advances, and then in resonance with the global florescence of Islamic militancy. But the latter contingency posed a problem, for though Belgrade consistently supported the Arab cause against Israel, it was quite apprehensive about any spillover of pan-Islamic allegiances into Yugoslavia's own delicate interethnic balance. In the spring of 1983, a number of Bosnian Muslims (including two clergymen) were tried and sentenced for illegal contacts with "reactionary" foreign imams and seditious activities. The authorities thus sought to differentiate between legitimate domestic ethnocultural group solidarity and unacceptable external political influences. For the time being, the issue was stabilized, but given the dynamics of ethnopolitical militancy in general and of Islam in particular, it was unlikely to have been solved definitively.[28]

An even more serious challenge to Yugoslav nationality policy was posed by the new assertiveness of the Kosovo Albanians, which not only rocked the delicate interrepublican balance of the federal system, but also undermined the comfortable traditional assumption that economic progress soothes ethnonational grievances.

Following severe riots in November 1968, Belgrade had acted quite vigorously to improve Kosovo's economy and to promote ethnic Albanians to positions of authority there. (In a classic yet nevertheless ironic pattern, these disturbances had probably been precipitated and "invited" by the lifting of Ranković's iron-handed sway two years earlier.) But these economic and administrative concessions by Belgrade assuaged tensions for only a decade until the province exploded still more violently in the spring of 1981. Even during the overtly nonviolent 1970s, the situation had remained strained, not only between the local Albanians and the governmental authorities, but also between the rapidly expanding and spreading Albanian provincial majority and the recoiling, hitherto dominant, Serb minority, which also regards Kosovo as sacred ancestral soil. Under these tense and volatile conditions, Belgrade had tried to walk a fine line, arresting troublemakers while conceding "legitimate" Albanian demands and restraining the Serb backlash.

But a series of rolling Albanian rampages between March and May 1981, which escalated to near-insurrectionary proportions as about 1,000 people were killed, indicated that Belgrade's delicate balancing act had failed. The Yugoslav army sent in Slovenian and Croatian (that is, ethnically neutral) units to impose a state of siege on Kosovo, virtually severing its communications with the rest of Yugoslavia and the outside world. Nevertheless, trains continued to be derailed, power stations and factories to be dynamited, and government buildings to be torched. The contagion spread to the Albanian minorities in neighboring Macedonia and Montenegro. Not until 1983 were public safety and public property once again secured in the Albanian-populated areas of Yugoslavia.

Overtly and ostensibly, the Albanians were demanding only the elevation of the Kosovo autonomous province to full and equal republican status, and not secession from Yugoslavia and annexation to Albania. Nevertheless, the Serbs alleged that the latter was their real hidden agenda, and Belgrade purported to detect the conspiratorial hand of Tirana behind the revolt. In any event, given the intensity of the Serbian (and, eventually, the Macedonian and Montenegran) backlash, Belgrade could no more concede the former than the latter scenario without suicidally alienating Yugoslavia's most numerous and central ethnonation, the Serbs. Yet the regime could also not simply crush and repudiate the Albanians without gravely compromising its proud Partisan heritage of being the transethnic reconciler of all the peoples of Yugoslavia. The upshot was a peculiar trade-off whereby the Albanians, though denied their political demand for republican status, were enabled to win in the demographic dimension, outbreeding and outsettling the beleaguered local Serbs, who steadily abandoned their ancient homesteads in Kosovo and migrated into core Serbia, where their tales resonanted to sympthetic anger. But the Albanians also remained poor, unemployed, and aggrieved. The trade-off was thus a time-buying palliative that contained the problem without solving it. And Belgrade remained understandably concerned lest this conflict reinfect other festering ethnic wounds.

It would be an error to believe that interethnic friction, in Yugoslavia or anywhere else, is directly "caused" by socioeconomic inequality between and among ethnic communities. By the same token, it would be an illusion to expect that correcting such inequality would necessarily solve the ethnonational "question." Nevertheless, there is no gainsaying that such inequality exacerbates the friction. And the Yugoslav record has been problematic. Despite a politico-ideological commitment to eventual equality and substantial developmental advances in the poorer ethnoregions (the south), the gap between them and the richer ones (the north) has been widening. Moreover, much of the impressive investment in the poor regions has gone into capital-intensive, not labor-intensive, extractive and heavy industries and infrastructures, thus leaving many of the social pathologies of underdevelopment (such as unemployment and apathy) relatively untouched and unimproved. Yugoslavia's continuing low productivity and chronic stagflation also impair its competitiveness in hard-currency countries.

Yet, its ominous fragilities acknowledged, the Yugoslav political system has also registered some real, if prosaic, recent successes. It remains hospitable toward and supportive of the cultural and educational instrumentalities and aspirations of all its component ethnonations. It continues to render to its citizens a wider range of civil freedom and a broader choice of consumer products than does any other Communist system. Despite dire predictions to the contrary, it weathered the death of Tito without general turmoil, without elite fragmentation, without institutional discontinuity, and without loss of effective political independence—in short, without the often prophesized "succession crisis." This suggests that it has retained adequate levels of legitimacy and of efficacy, despite the many disappointments since its heroic wartime phase, the grayness of Tito's collective heirs, and the defensive conservatism of the middle-level cadres. There is now a pervasive sense that systemic reform and rejuvenation are needed. Although there is no corresponding consensus on the direction, pace, initiation, scope, or control of such reforms, the fact that this system has several times in the past been capable of making quite profound changes without disintegrating renders plausible the expectation that its elite can once again make some hard decisions and master Yugoslavia's always delicate domestic stresses. But the great international roles of the Titoist years—as the solvent of Stalinism, as the shaker and then the balancer of the Communist world, as the energizer of the nonaligned camp, as the self-proclaimed innovator of a new type of politico-economic model with supposedly global significance—are beyond the capacity or even the aspiration of the Yugoslav system today.

6

A Precarious Stalemate

1

In the years following their countries' linked crises of 1956, the political trajectories of Poland's Władysław Gomułka and Hungary's János Kádár crossed each other. Whereas Kádár, who was originally loathed by his countrymen for having betrayed Imre Nagy and endorsed the Soviet crushing of the Hungarian Revolution, eventually earned their appreciation for initiating economic improvements and easing political constraints, Gomułka squandered the reservoirs of popularity and legitimation that had both sustained and been deepened by his return to power in the teeth of Soviet disapproval in October 1956.

A broad national consensus had enveloped Gomułka on the morrow of his resumption of power. Though ideologically anti-Communist and sentimentally anti-Russian, the Polish people understood that any challenge to their country's alignment with the Soviet Union, though it might be emotionally gratifying, was rationally precluded by the sheer facts of geography and power, the need for continued Soviet underwriting of Poland's postwar western and northern borders, and their country's economic dependence on the Soviet Union and the other countries of the Soviet bloc. Hence they were ready to mute their visceral skepticism about Gomułka's foreign policy and to acknowledge its correspondence to Polish *raison d'état*. They were even prepared to accept his insistence on the Communist party's continued monopolization of power, since this was, in the given circumstances of 1956, an inevitable corollary of Poland's membership in the Soviet bloc. But beyond these necessary accommodations to an ineluctable reality, the Poles were not convinced of the need for Gomułka's seemingly gratuitous domestic stringency, which soon surfaced. His ideological rigidity, his intolerance of dissent, his refusal to sweep unrepentant Stalinists and incorrigible mediocrities out of the middle rungs of the party apparatus, his failure to elevate professional integrity over partisan loyalty in the state bureaucracy, his crabbed puritanism, his seemingly principled preference for austerity over abundance, and hence his failure to sustain the hope—let alone achieve the reality—of a better life for the nation appeared to exceed the requirements of the Soviet alliance and even the demands of the Soviet leaders, who were soon signaling their endorsement of "goulash socialism" (more consumption and looser reins), as applied by Kádár to Hungary. Hence these traits eventually eroded Gomułka's originally vast popularity and blighted the high hopes that were placed in him in 1956.

Of course, this erosion of popular good will was not an overnight process, nor did Gomułka really renege on his own understanding of what had been achieved in 1956. He promptly reversed the collectivization of agriculture, permitted three-quarters of the arable land to revert to the independent peasants, and raised agricultural investments, thus rendering Poland the only state in the Soviet bloc whose agriculture is primarily in private ownership. He also curbed and punished police abuses, ended arbitrary political arrests, repatriated the Soviet military personnel who had been seconded to the Polish armed forces, acknowledged the patriotism of the London-loyal wartime resistance, permitted the reintroduction of religious instruction in the state schools (which had been unilaterally abolished in 1955, though sanctioned in the church–state agreement of April 14, 1950; see Chapter 3, section 2), authorized authentic discussions in parliamentary committees, and asserted Poland's right and determination to administer its own affairs in the context of a nonexploitative relationship with the Soviet Union. (In contrast to Tito, Gomułka modestly and prudently refrained from claiming any "exportable" general validity for these Polish deviations from Soviet norms.)

But in such matters as truly critical intellectual freedom and intraparty debate, Gomułka showed himself to be a restrictive Leninist and an anti-intellectual Khrushchevite to a degree not anticipated by the public. By October 1957, he had retightened censorship, closed the spunky youth newspaper *Po Prostu* (*Straight Talk*), and expelled its editorial board from the party. In the process of establishing his control over the party cadres—who, he insisted, "must be monolithic"—Gomułka dealt more severely with those whom he termed "revisionists" (the people who had elevated him to power) than with those whom he conceded to be "dogmatists" (the ex-Natolinites). His suspicion of the spontaneously formed workers' councils of 1956 was profound, and by the spring of 1958, they had been neutralized and resubjected to party control. On the execution of Imre Nagy in June 1958, Gomułka waited twelve days before commenting that though this retribution was "severe," it "settles the account with the counterrevolution that had occurred in Hungary." Some liberals appreciated the delicacy of the delay between the deed and this comment, but most Poles were ashamed. A year later, Gomułka forced the dismissal of the creative and probing young Marxist-humanist Leszek Kołakowski from the editorship of Poland's leading journal of philosophy (but did not jail him, in contrast to Tito's treatment of Djilas). By then, the popular enthusiasm for Gomułka had largely waned in direct proportion to such steady implementation of his Leninist conception of power.

Thus by the early 1960s, Gomułka's vivid repudiation of "revisionists" and "liberals," coming on top of his success in containing the Polish effervescence of 1956, had earned him the confidence of the formerly suspicious Soviet leadership to such a high degree that his erstwhile Natolinite rivals were marooned and neutralized. In their place, however, a new party faction arose to pressure, though not explicitly to challenge, him. Dubbed "the Partisans" and composed of relatively younger veterans of the wartime Communist underground apparatus, the faction's political values were a combination of chauvinism, militarism, and hostility to intellectuals, liberals, ethnic minorities, and "Muscovites." Its characteristic political style was marked by brutality, severity, and xenophobia. Led by a resourceful and energetic man who retained his *nom de guerre* of Mieczysław Moczar (then deputy minister and from 1964 minister of the interior), the Partisans were strongly seeded in the military and security apparatuses and the veterans' organization. Their attitude toward Gomułka was ambivalent, endorsing his increasingly emergent illiberalism but regretting his refusal to resort to nationalistic demagoguery and his related cordiality toward Moscow. He, in turn, sought to contain the Partisans with the more pragmatic fraternity of political, economic, and technocratic administrators, of whom the most promising was Edward Gierek, "boss" of the highly industrialized, seemingly efficient, and relatively prosperous southwestern region of Upper Silesia.

The Arab–Israeli Six-Day War in June 1967 further aggravated

Poland's domestic political malaise. To Gomułka's intense embarrassment, Polish public opinion empathized emphatically with Israel, largely because the Arab states were regarded as Soviet clients. To clear himself with Moscow, Gomułka not only severed diplomatic relations with Israel (as did all the other East Central European states except Romania), but also launched a year-long "anti-Zionist" drive that entailed a purge of the remaining Jews in Polish public life and of some Poles deemed to be too close to them. As Gomułka's personal political record had been free from anti-Semitism, it appears that his conduct at this juncture was being constrained by Soviet pressure from without and Partisan agitation from within Poland. Indeed, in March 1968, fearing that this "anti-Zionist" campaign was getting out of hand, Gomułka tentatively but vainly sought to curb it. Most of Poland's Jewish remnant as well as a sprinkling of non-Jewish liberals emigrated at this time, and the beneficiaries of the resultant mobility escalator in the political system were Moczar's Partisan cohort.

Hard on the heels of this affair came the general political crisis of 1968 in neighboring Czechoslovakia, which had a serious impact on Poland. An early spillover was Polish student unrest in March 1968 in response to Gomułka's prohibiting the production of a classical nineteenth-century drama sharply critical of the tsarist Russian control of Poland (*Forefathers' Eve*, by Adam Mickiewicz). The student disturbances and their suppression were not only the tail end of the "anti-Zionist" campaign, but also a harbinger of Poland's resonance to the "Czechoslovak spring," with which they shared an interesting characteristic: the student dissidents and their allied liberal reformers were cold-shouldered by the workers, who were aloof and skeptical of these efforts by the intelligentsia to "humanize" the Communist regime (see Chapter 5, section 5).

Gomułka's subsequent participation in the Soviet military overthrow of Alexander Dubček's Communist reform movement in Czechoslovakia in August 1968 was logically consistent. Indeed, there is a sustained line from the nature of his bargain with Khrushchev in October 1956, which facilitated his return to power in Poland, through his immediately subsequent restriction of "revisionists," the "anti-Zionist" campaign, the suppression of the students, to his willing role in snuffing out the "Czechoslovak spring" despite the contrary sentiments of the Polish people. By this time, the Sino-Soviet rift, the maverick behavior of Romania, the defection of Albania, the unreliability of Czechoslovakia, the brashness of Yugoslavia, and other developments in the Communist orbit had turned Gomułka's Poland into Moscow's most valued and pivotal ally. And Brezhnev showed his gratitude by endorsing Gomułka against his Partisan critics, even though in several ways their primitive style was more congenial to the Soviet apparat. Yet the crowning achievement of Gomułka's political career came from the opposite geographical direction, though it, too, may be regarded as a product of his unsentimental but sustained loyalty to the Soviet Union: the German Federal Republic's recognition on December 7, 1970, of the Oder-Neisse Line—that is, Poland's postwar territorial acquisitions at prewar Germany's expense. (This pivotal event was followed by the Vatican's final protocol arrangements for the full and formal incorporation of these western dioceses into the Polish Catholic church; see Chapter 3, section 2.)

Immediately on the heels of this triumph came Gomułka's downfall, brought about by his regime's blatant failures in the realms of economics and social justice. Poland's economy, marked by low consumption and declining capital productivity, had been limping for most of the 1960s—partly because of objective difficulties, such as the enormous reservoir of redundant labor accruing from high postwar birth rates; partly due to organizational distortions, such as excessive centralization; partly as a result of myopic policies, such as overinvestment in heavy industry; and partly thanks to the poor judgment of mediocre officials. To pay for important industrial equipment, Poland not only exported food at the expense of domestic consumption, but also paid its peasant-farmers so little for their products as to discourage their output. Ostensibly to correct these irrationalities, food prices were sharply increased and industrial wage scales radically revised by a decree of December 13, 1970, a week after the diplomatic triumph on the frontier issue. But the timing was so close to Christmas and the content of the decree so seemingly threatening to the real incomes of the already suspicious workers and demoralized housewives that it triggered strikes across the country as well as riots and mass demonstrations that lasted for five days in the Baltic port cities of Gdańsk, Gdynia, and Szczecin, whose inhabitants were particularly aware of higher living standards in Sweden, Denmark, and even East Germany. The suppression of this insurrection in the coastal cities required the use of regular army forces and entailed considerable bloodshed and demoralization. But just as the students had been left in the lurch by the workers in 1968, so the intelligentsia did not actively join the workers in 1970. Hence these insurrectionary events did not reach the scale of a comprehensive revolution.

Nevertheless, this tragedy finalized the erosion of Gomułka's authority among the masses, the party cadres, the ruling elite, and probably even the leaders in Moscow. Stricken by a mild cerebral stroke on December 18, he was replaced as first secretary by Gierek two days later. The incendiary decree of December 13 was rescinded in mid-February, effective March 1, 1971, symbolizing a defensive victory of the workers over the regime. Though Gomułka's last years in power had been tarnished by errors and repressions, he merits retrospective credit for his personal probity, his political courage during the Stalinist purge years 1948 to 1954, his dignified patriotism in 1956, and his consistent aversion to systematic mass terror. The only Polish Communist ever to have been, albeit briefly, an authentic national hero, he died, forgotten, on September 1, 1982, at age seventy-seven.

This is a suitable moment for a brief discussion of the political problems that result from consumer-price increases in Communist-type administered economies, an issue that toppled Gomułka in 1970, would over the next decade savage his heir, Gierek, and troubles similar regimes in other countries. In Communist systems, administrative and political inertia tends to hold prices, wages, and work norms relatively stable for extended periods. Simultaneously, the steady introduction of labor-saving machinery tends to facilitate the gradual overfulfillment of the work norms, leading to screened but real increases in take-home wages. These processes are incremental. Periodically, however, the regimes feel compelled to close the resultant creeping gap between stable prices and wage norms, on the one hand, and rising incomes, on the other hand, by way of a sharp increase in prices and/or a tightening of work norms. But this is a nonincremental, precipitous event likely to provoke anger among consumers and workers, which may erupt into disorder and may tempt one faction of the regime into demagogic opportunism while another faction tries to salvage the situation by "toughening-out" the discontent.[1] All these scenarios were played out in Poland during the 1970s and 1980s.

Gierek's basic problem was the same as Gomułka's—how to reconcile Communist rule with national aspirations—but his style in approaching it was the reverse of his predecessor's. Puritanical frugality was replaced by cavalier extravagance. Instead of scolding the Polish people for supposedly living beyond their means, Gierek catered to their craving for consumer goods. Now that the German irredentist threat had ceased to be plausible, with Bonn's recognition of Poland's frontiers, the rhetoric of sacrifice and siege could yield to images of indulgence and expansiveness. And in contrast to the aloof and self-righteous

Gomułka, Gierek undertook a series of personal dialogues with the restless workers, whose scattered slowdowns and strikes persisted through the early months of 1971, and dismissed some officials whom they deemed responsible for the December 1970 tragedy.

In addition to his populistic touch for grass-roots contact with the masses, Gierek was skilled at power politics within the elite. During 1971, he deftly neutralized his rival Moczar, first by sidelining or coopting his key supporters and then by "kicking him upstairs" into a rather powerless sinecure. Other holdovers from the Gomułka era whom Gierek perceived as potential rivals or embarrassments were also relegated to political oblivion and replaced by his own peers and protégés. And within this cohort of his own associates, Gierek initially practiced a more consultative style of decision making than had characterized Gomułka's domineering stance toward his entourage. Finally, Gierek resumed the positive engagement with the Roman Catholic hierarchy that Gomułka had soured since the early 1960s.

Eager to help him stabilize Poland in the era of détente, Moscow granted Gierek large credits for the importation of Soviet grain, oil, cement, and housing components, as well as permission to raise additional capital in the West, where banks were soon conveniently awash with excess liquidity accruing from the exponential oil-price rises of 1973 and the subsequent recycling of these petrodollars. Thus with Western banks and governments craving to lend this money to plausible borrowers, Poland's foreign indebtedness soon reached imprudent levels because, apart from some initial concessions to the private-sector peasantry, no serious, deep structural reform of the state-owned industrial sector, such as might function as a lasting inducement to greater productivity, was undertaken. In effect, the massive Soviet and Western loans that flowed into Poland during the 1970s were largely wasted, for they subsidized only an orgy of consumerism, of corruption, and of overoptimistic, grandiose, uncoordinated, and hence senseless investments. Gierek's seemingly impressive political and public-relations performance was built on economic sand. Hence like the pharoah in Genesis, he would see his initial series of good years swallowed by a hard and troubled second half of the decade.

Attempting to correct the budgetary distortions caused by his profligate borrowing-and-spending policies (but whose ultimate causes ran much deeper), Gierek felt himself driven to the same fateful step as had tripped Gomułka: a steep and unanticipated increase in basic food prices, announced abruptly on June 24, 1976. His failure to consult the public in advance was an ominous deviation from his earlier style, signaling a peculiar mixture of hubris and desperation. This time, the reaction was sharper among the workers in the industrial cities of central Poland than among those on the Baltic coast. Their uproar forced the rescinding of the announced price increases the next day. Meanwhile, hundreds of demonstrators were arrested, scores maltreated and beaten by the police, and many eventually sentenced to prison. Though Gierek survived in office, the blow to his authority and prestige was severe; the fragile fabric of reconciliation between the regime and the society that he had worked hard to weave in the preceding six years was again rent asunder.

The already brittle legitimacy of Gierek's regime was then dealt another damaging blow when the Roman Catholic College of Cardinals elected Karol Cardinal Wojtyła, the archbishop of Kraków, as Pope John Paul II on October 16, 1978. This event released a surge of religious nationalism, self-confidence, and euphoria throughout Polish society, compared with which the disarray, isolation, indecisiveness, and defensiveness of the Communist government made a sorry spectacle. And the new pope's triumphal nine-day visit to his homeland the next June further highlighted this stark contrast between a vibrant people and a sclerotic political system. Worse yet for the regime, the papal election and visit coincided with several politically evocative national anniversaries: the sixtieth anniversary of restored Polish independence (1918), the fifty-eighth anniversary of the Polish-Soviet War (1920), the fortieth anniversary of the German-Soviet partition of Poland (1939), and the thirty-fifth anniversaries of the Warsaw insurrection and the founding of the Communist state (1944).

With his room for political maneuver shrinking ever more tightly, Gierek desperately fought for survival by trying to retain access to Western capital markets while switching to retrenchment at home. But neither leg of this dual stance would stand up under pressure. Between 1975 and the summer of 1980, Poland's hard currency debt to the West tripled, from $7.4 billion to over $21 billion. By the latter date, much of this stood exposed as "bad debt," and Poland's credit rating was approaching exhaustion, a condition aggravated by the simultaneous ending of the American–Soviet détente. And the belated effort at domestic retrenchment provoked confusion and frustration as factory managers found themselves trapped between central directives and the workers' scorn. The whole economic system was becoming anarchic as multiple shortages and bottlenecks disrupted the continuity of production. The bubble finally burst on July 1, 1980, when the Communist rulers resorted for the third time to the gambit that had already backfired twice before—announcing meat-price increases without preparing or consulting the public. This decision seemed so perversely suicidal that some observers believed it to be a deliberate provocation, intended to elicit protests and disorders that could then be conveniently crushed, thereby facilitating a reversion to a comprehensively repressive system. But while there was indeed a hard-line faction in the party that dreamed of "restoring discipline," this suspicion underestimates the sheer bewildered incompetence of the Gierek regime in its death throes.

This time, the hitherto divided participants in previous protests—workers, students, intellectuals, housewives, even peasants—united in a solid phalanx, whose spearhead was, as in 1970, the Baltic shipyard workers. Much of the credit for this solidarity of the protest coalition belongs to the Committee for the Defense of Workers, or KOR (Komitet Obrony Robotników), which had been organized in the summer of 1976 by dissident intellectuals and professionals to provide legal assistance to the workers being persecuted for having taken part in the strikes and disturbances in June, material assistance to their families, and publicity about the authorities' behavior. Multigenerational, ideologically diverse but committed to an amalgam of democratic socialism, patriotism, and civil libertarianism, KOR was the most vivid but scarcely the only expression of a burgeoning process that was termed the "self-organization of Polish society" outside, parallel to, and in the teeth of an incompetent political system.[2] Several million citizens were eventually drawn into this process, which, though secular, enjoyed the benevolent regard of the Roman Catholic church. The church, in turn, under the astute leadership of Cardinal Wyszyński, positioned itself as a political reality but ostensibly not a political actor, as being "in the world but not of it." Hence its protection of the "self-organization" process was mutually beneficial. And the Communist party, which had still been able to lead and control the great national catharsis in 1956, was in 1980 reduced to floundering marginality by the symbiosis of the church and the "self-organization" movement.

The disciplined wave of nonviolent strikes that rolled over Poland in July and August 1980 was energized not only by all the grievances to which allusion has been made—the regime's remoteness, exclusiveness, arbitrariness, unaccountability, incompetence, repressiveness, and illegitimacy—but also by a profound revulsion against the particular failings of the regime's current set of leaders and functionaries—their contempt for social justice and their corruption, careerism, arrogance, frivolity, and mediocrity. And the strikers felt secure in their own superiority to the regime in these several moral and political dimensions. Their self-confidence was further enhanced by the blatant contrast between their own cohesion and the factionalism that was now exposed in the regime. Hence the appropriateness of the name Solidarity (*Solidarność*), which they gave to the organizational expression of their movement.

On August 31, 1980, Lech Wałęsa, leader of the Gdańsk, Gdynia, and Sopot strikers, signed an agreement with the government by which the workers won not only their demand for immediate wage increases and price-rollbacks, but also the legal right to strike and, even more portentously, the right to organize independent trade unions, free from party control—a concession unprecedented in any Communist society. The agreement also provided that the state radio system broadcast Catholic Sunday Mass, that political prisoners be released, that the dismissals and persecutions of the strikers of 1970 and 1976 be reexamined, that competence rather than party affiliation be the criterion for the designation of managerial personnel, that the nation be informed about its economic situation, that health, social security, and housing services be improved, and that the communications media be opened to a variety of views and opinions. Supplemented by similar agreements at other strike locations, this accord not only disoriented the Polish Communist party—now pitilessly exposed as representing not the working class but only its own bureaucracy—but also seriously alarmed all the other ruling Communist parties.

For the next sixteen months, Polish political developments were circumscribed by six interlocked facts, circumstances, and contingencies: (1) The language of the Gdańsk agreement (and of analogous accords elsewhere) was in places imprecise enough to permit contrasting interpretations of its timing and scale. (2) The economy was in far worse shape than either party to the negotiations had realized, rendering some of the agreement's clauses illusory. (3) The party split into a hard-line faction, convinced that the imperatives of sheer survival required that the agreement be nullified, and a moderate group, prepared to live with it in order to avoid even more damaging civil unrest. (4) Solidarity, whose conception of democracy entailed a rather loose, participatory, consensual style of decision making, was similarly strained between pragmatists searching for a *modus vivendi* with the residual Communist apparatus and activists determined to turn the screws of public pressure ever more tightly against the hated and now exposed system. (Both wings were, however, agreed that Solidarity should not make a revolutionary bid for governmental power or demand Poland's repudiation of its alliance commitments to the Soviet Union and the WTO.) (5) Moscow and the other bloc regimes deemed the agreement's political concessions (not necessarily its economic or social clauses) to be unacceptable. (6) The Roman Catholic hierarchy, though endorsing the "dignity of labor" aspects of Solidarity's quest, was also recommending tactical restraint and strategic moderation lest Poland again lose its sovereignty. ("We must live with what we have.") This matrix of issues and pressures was bound to generate much furious maneuvering by all the actors.

On September 6, the party replaced the universally execrated Gierek as its first secretary with Stanisław Kania, an apparatchik whose career had been in the crucial areas of agriculture, personnel assignment and monitoring, security, and party–church relations. Pleading for patience, partnership, Polish discipline, and Soviet understanding in an environment where all these were in short supply—a polarizing environment of bureaucratic obstruction and hard-line truculence by middle-level and regional apparatchiks, of selective warning strikes by Solidarity, of disillusioned bitterness among the party rank and file, and of ever intensifying Soviet insistence that the Solidarity abscess be lanced lest the Polish disease infect the entire bloc—Kania was unable to master the crisis. On February 9, 1981, the drain in domestic and foreign confidence obliged him to designate as his government's prime minister General Wojciech Jaruzelski (defense minister since 1968), who thereafter steadily eclipsed Kania until, on October 18, Jaruzelski replaced Kania as the party's first secretary, combining the highest governmental and party positions in his own person.

Though a member of the party's Politburo throughout the 1970s, Jaruzelski was also a thoroughly professional military officer. Hence the fact that the Polish Communist party now found it necessary first to let him head the government while retaining control of the armed forces as defense minister, then to make him its own leader, and finally to let him seed other military officers into the most pivotal positions of the governmental apparatus is a devastating judgment on the nullity of that party's politicians and administrators. It is also an implicit embarrassment to the Leninist heritage and ideology of the Soviet leadership, which has always prudently nipped any incipient or suspected "Bonapartism" in the bud. But in the Poland of the 1980s, the Soviets appear to have reconciled themselves to the absence of a viable alternative. In contrast to the discredited party, the Polish army had remained a respected institution, and Jaruzelski was reputed to have opposed its use against the striking workers in 1970 and to have vetoed its use against them in 1976. Yet his career also expressed external loyalty to the Soviet alliance. He thus seemed to be Poland's last chance to resolve its crisis short of civil war and/or Soviet invasion.

Jaruzelski began his prime ministership with a rather imaginative call for ninety strike-free days while his government would start to stabilize the economy and rationalize public policy and administration. Inviting "comprehensive social dialogue," he projected an image of reasonableness and determination, and Wałęsa was ready to meet him halfway as a partner in the task of national renovation. But neither leader could persuade his followers to credit the other's bona fides. On the one side, the radical rhetoric of some Solidarity figures stoked the anxiety of the regime's apparatus that the new organization's political ambitions were limitless and essentially subversive. On the other side, the chronic procrastination of officialdom in implementing the accords of the summer of 1980, and the occasional brutality of the provincial militia and security services, fed the suspicion of Solidarity and the "self-organization" movement that the rulers' professions of turning over a new leaf were fraudulent. Meanwhile, continuing shortages of food and other essentials and the interminable, exhausting lines necessitated by these shortages took a heavy toll on the citizens' spirits. A pervasive mood of suspicion, irritability, and bitterness lamed Jaruzelski's and Wałęsa's groping toward dialogue and negotiation. Simultaneously, the Soviet complaints about Poland's "antisocialist and anti-Soviet bacchanalia" were becoming ever more minatory and were backed by massive military exercises. Particular outrage was expressed by Moscow over a mildly worded but nonetheless imprudent message of fraternity addressed by Solidarity on September 8 to the workers of the other East Central European countries and of the Soviet Union "who have decided to enter the difficult struggle for a free and independent labor movement; we hope that our representatives will soon be able to meet each other."[3]

Under these antagonistic yet mutually reinforcing pressures of a deteriorating domestic situation and a mounting external threat, Jaruzelski proclaimed martial law ("a state of war") on December 13, 1981, established the Military Council of National Salvation to administer the country (with no explicit role for the party), suspended Solidarity, and interned thousands of its leaders and activists (and others opposed to the regime). At one level, he was correct in believing that the nation was now tired of turmoil, dubious about Solidarity's wildcat tactics, and ready for a return to order and work. At a deeper level, however, he underestimated the moral and political cost of so blatantly reneging on the implicit social contract between the regime and the people that was symbolized in the accords of the summer of 1980. Though decapitated and subsequently outlawed, Solidarity survived underground and reconsolidated the somewhat shaken sympathy of the general public. By the mid-1980s, Jaruzelski was politically secure enough to end martial law, release many political prisoners, ease the censorship, and devolve the prime ministership onto another person while retaining the party leadership and exchanging the Defense Ministry for the chairmanship of the National Defense Committee (through which he reserved to himself ultimate control of the armed forces), but he failed to regain the trust of the Polish people. Nor did he satiate the even more repressive itches of the party's and the security police's unappeased hard-liners—itches that were starkly exposed by the brutal murder by the police of Father Jerzy Popiełuszko, a vocal opposition priest, in October 1984. And the economy remained in shambles. Poland, in short, was in the grip of a civil–political stalemate. On the one hand, the regime could not obtain legitimation from a defeated, resentful, yet contemptuous and still defiantly "self-organizing" society that stubbornly insisted (in Adam Michnik's resonant phrase) on "living as if we were free." On the other hand, this society could neither obtain access to the forums of political decision making nor convert its potency into power.

But in politics, unlike chess, stalemate is not a final condition, nor does it end the game.

2

Whereas Poland's economic miasma during the 1970s first had helped to precipitate Solidarity's challenge to the regime but then had encumbered the political thrust of that challenge, in Hungary during the same decade, economic recovery and progress served to stabilize and enhance the acceptability of the Communist regime. But before this Hungarian economic success could be launched, the nation's and the working class's political will and political resistance had to be broken. And here, in the aftermath of the 1956 revolution, János Kádár and his Soviet sponsors had proceeded more ferociously and terroristically than Jaruzelski did in Poland after 1981. Moscow also subsidized Kádár far more generously with economic aid. In short, in Hungary the Soviets finally learned the lesson of their several half-measures during the years 1953 to 1956 (see Chapter 5, section 3). Hence they and their local clients proceeded with the postrevolutionary "normalization" in two delineated, sequential steps: (1) crush the society into utter defeat and submission; and (2) then gratify and tame it with economic rewards and material satisfactions. Parenthetically, Jaruzelski has not assayed either of these two phases in post-1981 Poland.[4]

During and immediately after the 1956 revolution, over 200,000 Hungarians emigrated to the West, tens of thousands were imprisoned, about 2,000 were executed, and others were reportedly deported to the Soviet Union. All the workers' councils, revolutionary committees, intelligentsia organizations, and political parties that had been born or revived during the revolution were dissolved or emasculated. Thus Kádár was early and radically disencumbered of his most recalcitrant resisters and the most committed bearers of a would-be political counterculture. In 1959, he followed up this political cleansing with a harsh reimposition of agricultural collectivization (which had unravelled in the revolution), a process completed by 1962. Having reached this milestone, Kádár shrewdly veered, announcing the end of domestic class struggle and inviting the now beaten society to ally itself with his regime in pursuing the uncontroversial goal of economic development and modernization, simultaneously offering to it relief from further terror and coercion. ("Those who are not against us are [deemed to be] with us.") Political prisoners were amnestied, police terror was curbed, class bias in access to higher education was eased, the contributions of nonparty experts were welcomed, the cultural arena was rendered more tolerant, and the residual supporters of Mátyás Rákosi were swept out. In effect, Kádár was bidding for pragmatic acceptance by Hungarian society of his *government* via the route of economic and scientific progress while refraining from pressing for ideological legitimation of the Communist *system*. Indeed, in later years, he went even further, implicity hinting that such societal support for his government and its policies would even enable him to protect the citizens from the potential repressiveness of the system itself.[5] And the Hungarian people as well as its intelligentsia responded to Kádár's overtures; exhaused by the upheavals of revolution, repression, and the enervating posture of silent "inner emigration," they wanted to return to work and to "normalcy," even at the price of painful self-censorship.

Kádár's license, as it were, for his new ameliorative course was Khrushchev's resumption of his attack on Stalin's heritage at the Twenty-second Congress of the Soviet Communist party in October 1961, an attack that Kádár welcomed as emphatically as Rákosi had resisted the first round after the Twentieth Congress in February 1956. And Kádár held his course even after the Soviet invasion of Czechoslovakia in August 1968, which might have cowed a less committed reformer. Of course, the fact that Moscow perceived him by then as a veteran, reliable leader meant that Kádár enjoyed a longer leash of Soviet forbearance than had been conceded to the less familiar and less experienced Dubček in Czechoslovakia.

The logic of Kádár's stance led eventually to an array of decentralizing reforms, market-oriented innovations, privatizing entrepreneurial adaptations, and civil-libertarian concessions that cumulatively swept Hungary into the role of a pioneer of successful Communist reform—a role that was indeed historically ironic, given the background of the Rákosi era and of Kádár's own behavior in 1956. It was also a role that he never trumpeted as a model for others (à la Tito), lest Moscow be irritated.[6] Of course, none of these reforms impinged on the two great political taboos that were briefly violated in 1956: the Communist party's monopoly of political power, which it would neither share nor submit to a free electoral test, and the country's anchorage in the Soviet bloc. Nor did the reforms prove to be a panacea for social stress and periodic economic sluggishness, though they vastly improved the supply, range, and quality of consumer goods and services as well as of food. "Goulash socialism" was extended into "refrigerator socialism," "car socialism," and, for a large minority, even "weekend-cottage socialism." The reforms' capstone was the New Economic Mechanism (NEM), emplaced in 1968, briefly degraded between 1974 and 1978, and renovated in 1979.

In contrast to the hesitant, piecemeal, and "controlled" manner in which the Soviet leadership occasionally tinkered with some economic reforms in those years, the NEM was a multifaceted, multidirectional, wholesale, aggressive, structural overhaul of the Hungarian economy. Among its more interesting novelties were initiating autonomous self-management of the collective farms in regard to both production and marketing decisions, together with actively encouraging supplementary private-plot cultivation; developing sophisticated agricultural industries; breaking up monoply enterprises; curtailing subsidies (except export subsidies); linking prices to the world market via semi-convertibility through multiple exchange rates; authorizing teams of workers to produce independently in the state-owned plants after regular working hours; introducing a series of other productivity incentives; legalizing private artisanal, retail, and service activity; and substituting economic regulators for compulsory directives in the "socialist," or state-owned, sector of the economy, which remained, of course, the dominant sector.[7] The NEM generated a national economy that was, on the one hand, complex, innovative, and relatively efficient but, on the other hand, highly dependent on foreign trade and hence vulnerable not only to general fluctuations in the world markets, but especially to periodic administrative escalations in the prices of imported Soviet raw materials and energy resources (Hungary being resource-poor). For a regime that based itself so heavily on gratifying its public through prosperity rather than controlling it coercively or normatively (ideologically), such vulnerability, though unavoidable, was quite worrisome.

Though Kádár's relative pragmatism and reformism entailed a substantial degree of ideological agnosticism, social careerism, and political opportunism among the party cadres, the technologically qualified nonparty intelligentsia, and the new middle strata of professionals and managers, Hungarian society was not depoliticized to anything like the extent that was often predicted in the 1970s. Nationalist, neo-Marxist, and anti-Marxist streams of strong political sentiment remained readily available to tap for such issues (usually embarrassing to the regime) as the survival of poverty and its ancillary social pathologies in Communist Hungary, the fate of the Magyar minorities in neighboring Communist countries, environmental degradation, European denuclearization, the rehabilitation of Imre Nagy and his fellow martyrs of 1956, and a neo-egalitarian backlash against the socioeconomic differentiations and restratifications accruing from the NEM. Though sporadically harassed, the dissidents who addressed these issues and themes were not systematically persecuted, as the Kádár regime was reluctant to submit its relative popularity to the hazardous test of systemic legitimation. Its policy toward intellectual dissent became, by East Central European standards, relatively tolerant. Indeed, on the issue of the fate of the Magyars in neighboring Romania, the regime eventually identified with the national outrage. In a sense, Kádár's choices were the reverse of Nicolae Ceauşescu's in Romania, who opted for a stance of unpopular domestic Stalinism characterized by repression, austerity, hyper-investment, and insistence on systemic legitimation, balanced by truculent nationalism and external defiance of the Soviet Union.

Over the course of three decades following his coming to power in

1956, Kádár transformed his reputation from that of a Soviet lackey who betrayed the revolution into that of a genial sage who was trusted by Moscow for his loyalty and for keeping his people quiescent, respected in the West for his economic pragmatism and political moderation, and valued by his own nation for alleviating the rigors of Communism and keeping Soviet pressure at bay. All three perspectives contained a mix of much truth and some humbug. Through the political and biological attrition of most others, Kádár also achieved the dignified status of an elder statesman of Communist East Central Europe, whose power was more consensually based than that of any other leader in the area. Yet his standing was highly personal and could not be automatically transferred intact to a successor. That successor was designated on May 22, 1988, when Kádár's own protégés, judging that at the age of seventy-six he had lost his political acumen, created for him the honorific sinecure of party chairman, and replaced him as first secretary with the energetic Károly Grósz, born in 1930.

3

Czechoslovakia under Gustáv Husák made a dour contrast with Kádár's buoyant Hungary. Though the two leaders had similar early credentials insofar as both had been wartime "local underground" Communists who had then fallen victim to the Stalinist terror of the early 1950s and had later been elevated to power under Soviet auspices, their stances and policies thereafter diverged emphatically. Rigid where Kádár was flexible, Husák consistently preferred dogmatic prescriptions to experimental risks. This reflected both his personal predilection as well as the fact that his Soviet sponsor was the equally conservative Brezhnev rather than Kádár's more elastic Khrushchev. The difference may also have been nurtured by the circumstance that the Czechoslovak resistance to the Soviet invasion was less militant, less bloody, and less cathartic than the Hungarian resistance had been; hence Husák may have felt a lesser need and craving than Kádár to stake out a fresh political position that would break emphatically with the past—both the immediate past and the harsher past that had preceded it. After coming to power, he did not punish the reformers as ferociously as Kádár did the revolutionaries (though he eased them out of public life and had them badgered and demeaned), but he also did not eventually implement much of their program and many of their demands, as Kádár did after 1962. Where

Kádár and Khrushchev had inflicted radical surgery followed by vigorous rehabilitation in Hungary, Husák and Brezhnev opted for medicinal dosages in Czechoslovakia. They numbed rather than terrorized. The net effect was that two and three decades after their respective convulsions and afflictions of 1968 and 1956, Czechoslovak society was more inward, lethargic, and slack and was characterized by lower morale than was Hungarian society. Whereas Hungary's regime ebulliently welcomed Western contacts and openings, Czechoslovakia's sought to restrict them and (until the ascendance of Mikhail S. Gorbachev) never probed the possibility of expanding its room to maneuver vis-à-vis the Soviet Union. While Kádár assumed that his people had came around to endorsing him and his policies, Husák—with good reason—made no such assumption.[8]

For a long interval, subsidized Soviet oil sales cushioned Husák's regime from the impact of escalating world energy prices. This permitted the Czechoslovak economy to perform well until the mid-1970s. Though capital investment continued to rise thereafter, productivity, consumption, and real wages declined. Moreover, the quality of Czechoslovak products deteriorated despite the continued infusion of large investments, leading to the loss of markets in the Western and developing worlds and to sharp Soviet warnings of impending similar losses there. Such stagnating performance reflected, of course, low morale and fatigued leadership. Husák's regime lacked both the vigor of true totalitarianism and the élan of real reformism. That is why it felt so threatened by the Solidarity effervescence in neighboring Poland in 1980 and 1981, when, in a role reversal from 1968, Prague was urging Moscow to stifle the alleged "counterrevolution" in Warsaw. And its sense of insecurity lured the Husák regime into a bunker posture of shunning economic reforms—quite in contrast not only to Kádár's Hungary, but also to Todor Zhivkov's Bulgaria, Erich Honecker's East Germany, and even Gorbachev's Soviet Union.

By the mid-1980s, the Prague government had at last decided to reduce investment in order to protect consumption, but still without structural reforms. This was basically and characteristically a defensive, preventive political decision that failed to address the deeper flaws of the economy, including its technological deterioration and its related tendency to squander energy and materials, swell pollution, degrade the environment, and harm the public's health. A significant but misguided substitute for corrective reforms was the regime's turning a blind eye to a thriving underground economy, fueled by embezzlement, theft, corruption, bribery, absenteeism, and entrepreneurship. This acquiescence sustained surprisingly high consumption and diluted political tension, thereby contributing to stability, but at the long-run cost of rampant cynicism, eroding social ethics, and aggravating social malaise. It also conflicted with Soviet leader Gorbachev's expectations about how "socialism" should administer itself. And it was ultimately a bind, not a solution, for it mistook stability for viability.[9]

Politically related to the regime's acceptance of the underground economy with its consumerist ethos (albeit on a somewhat different moral plane) was its toleration of a degree of privatization in daily life. Citizens who refrained from political involvement, sought no sensitive data, asked no embarrassing questions, disseminated no awkward information, flaunted no countercultural symbols, meddled in no public affairs, participated mechanically in the annual May Day rally, and voted reflexively for the party's slate of candidates could ordinarily anticipate being allowed to live their depoliticized and deideologized lives in peace, albeit as internal emigrants in their own country. Like the illegal economy, this realm of social privacy functioned as a political safety valve.[10]

Alas, the Husák regime's toleration of a certain degree of apolitical privacy did not extend to a more elevated and principled respect for civil liberties. Political trials, blacklists, and harassment of dissidents were quite pervasive and perversive, though the repugnant compulsory public self-derogations and self-humiliations of the 1950s were avoided. For the first few years after the Soviet invasion in August 1968 and its blighting of that year's "Czechoslovak spring," the public's disillusionment and hopelessness were so deep as to discourage dissident activity. By the mid-1970s, this depression had eased and a dissenting "typewriter culture," also termed the "parallel culture," began to emerge, analogous to but far weaker than the contemporaneous "self-organization" phenomenon in Polish society (see section 1). Its main organizational expressions, Charter 77 (founded in January 1977) and the Committee for the Defense of the Unjustly Persecuted (formed in 1978 and termed VONS from the initials of its Czech name), sought to anchor themselves in the Helsinki Final Act of 1975 and in Czechoslovakia's formal constitution. Calling on the regime to abide by its own laws and international commitments, they worked for civil and human rights while denying that they wished to play an opposition political role or even to offer a reformist social program. But this asserted self-restraint availed little, for the governmental authorities, subscribing to much more sweeping definitions of the terms *opposition* and *political*, persecuted the two groups, confiscated their materials, imprisoned some of their members, deprived their children of access to higher education, forced many into menial work, drove others into emigration, and intimidated still others into silence. The surviving Charter 77 activists retorted bravely by broadening their critique from the juridical and moral point of civil rights to the issues of economic decay, youth culture (defending rock and jazz music), and environmental degradation—to which there was an unexpectedly strong public resonance. Nevertheless, despite the sympathy of many ordinary people, Charter 77 and VONS remained weak and small, ever harrowed by the efficient police and largely confined to Czechs that is, sans Slovaks. It was symptomatic that whereas Polish underground publications in the 1970s and 1980s were often physically handsome and finely produced, the Czechoslovak ones—though equal and occasionally superior to the Polish materials in the intellectual and aesthetic quality of their contents—were typed, weakly bound, and distributed irregularly. If smuggled abroad, the Czechoslovak writings were sometimes republished professionally, but only a trickle ever found their way back into the country, where they were largely unknown.[11]

Nevertheless, the likelihood that despite their fragile and precarious condition, Charter 77 and VONS were the tip of a vast submerged iceberg of popular frustration and disaffection was suggested by the palpably negative mass reaction of Czechoslovaks to the deployment of Soviet nuclear missiles in their country toward the end of 1983. This event touched raw emotions and aroused deep fears. Even the industrial workers, who had remained largely aloof from the intelligentsia's dissent, subscribed to petitions protesting against the deployment of nuclear weapons "anywhere in the world"—petitions that deliberately shunned the conventional Communist differentiation between supposedly "peace-enhancing" Soviet and allegedly "aggressive" American weaponry.

The internal politics of the Husák government were characterized by a high degree of personnel and policy stability. Some natural deaths apart, the team that took power in 1969 under Soviet aegis remained by and large unchanged until the aged and ailing Husák (born in 1913) vacated the party leadership, while retaining the state presidency, on December 17, 1987. It had become an ossified oligarchy. To the extent that it was marked by any internal strains, they were between conservatives and even more reactionary hard-liners; innovative, moderate, and experimental Communists were absent from its ranks. Its operational code was "no surprises and no innovations." Yet for all its apparent stability and manifest control, this government projected an aura of perpetual uneasiness, defensiveness, insecurity, and irritability, exemplified by its habit of impounding its own statistics, suppressing its own surveys, and treating as utterly subversive the suggestion that it heed its own constitution and legislation. It was as though the Husák regime had anticipated its critics and preempted its dissidents in judging Communism in Czechoslovakia to be simply unreformable. Eventually, this judgment was bound to be shared by the dissenters and then by the general society. This development was an ominous "wintery" retrogression from the hopeful visions still entertained during the political "spring" of 1968. It also signaled a general sociopolitical deadlock, different from but no less grave than the Polish stalemate. And just as Antonín Novotný had dragged his heels against Khrushchev's de-Stalinization (see Chapter 5, section 5), so Husák three decades later resisted Gorbachev's suggestions to thaw the deadlock through *glasnost* ("political openness") and *perestroika* ("economic restructuring"). With good reason, for these new winds from Moscow, if steady and sustained, boded ill for its Prague satraps. In this context, the elevation of the Czech Miloš Jakeš (born in 1922) to the party leadership at the close of 1987 was, at best, enigmatic, his main achievement having been to orchestrate the purge of the supporters of Dubček after 1968.

4

Bulgaria was the only Communist-ruled country in East Central Europe whose regime never succumbed to the temptation to challenge Soviet hegemony, to deviate from Soviet directives, or even to test Soviet tolerances in any policy dimension—political, economic, or cultural, domestic or foreign.[12] Only once was this seemingly flawless Soviet–Bulgarian congruity deliberately ruffled by Bulgarian political actors, who were promptly squelched. The occasion was the puncturing in April 1965 of an allegedly "Maoist" military and political conspiracy to alter the government's total alignment with the Soviet Union. Given the gravity of the charged offense, the conspirators received surprisingly light sentences after their leader, a member of the party's Central Committee, committed suicide. Perhaps the mildness of the sentences suggests that the implicit nationalism of the plotters struck a wider responsive chord. Rumor had it that behind the official allegation of pro-Maoism there lurked screened internal party enmities stemming from World War II. Be that as it may, the party apparatus in the Vratsa region of northwestern Bulgaria, where the plotters had operated as wartime guerrillas, was purged after the unraveling of the conspiracy. Rumor also alluded to five additional failed plots during the 1960s but never imputed an anti-Soviet dimension to them. Perhaps such rumors were a projection of the fact that military *Putsches* had long been an established tradition in Bulgarian political life.

The absence of overt strains and crises between Bulgaria and the Soviet Union was not due to any culturally conditioned servility or historically ordained sentimentality on the part of the people or rulers of the smaller state. Indeed, the interwar tradition of the Bulgarian Communist party, with its oft-shown volitional contempt for reality, suggests a spiritual affinity for the "heroic," "forward-leaping" style of Maoist China.[13] Rather, Communist Bulgaria's sustained allegiance to the Soviet Union reflected a real complementarity of economic interests and developmental strategies. After Stalin's death in March 1953, his insistence on uniform priority of heavy industry throughout the bloc eventually yielded to a more modulated and variegated Soviet approach that in time permitted and even encouraged Bulgaria to concentrate on its real marginal advantages—light industry, agricultural industry, electronics, and tourism. This neatly dovetailed its economic interests with Soviet needs and lubricated its political loyalty, in return for which it received extensive and sustained Soviet capital assistance. (Of course, the earlier investments in heavy industry, for which Bulgaria lacked adequate and suitable resources, had built a complex of plants and mills that remained an albatross around the economy's neck.)

Our earlier narrative took Bulgaria through the Communist seizure of power, executed with special brutality and resisted with exceptional temerity, and the immediately subsequent intraparty purge (see Chapter 3, section 7; Chapter 4, section 3). In rapid sequence, two natural deaths bracketing one political execution promptly eliminated the trio of Communist leaders who had dominated the party during its years in the political wilderness and underground and then during its seizure of power: Georgi Dimitrov (July 2, 1949), Traicho Kostov (December 16, 1949), and Vasil Kolarov (January 23, 1950). The immediate beneficiary of the resultant fluidity was Dimitrov's brother-in-law, Vŭlko Chervenkov, who had lived in the Soviet Union from 1925 to 1946, where he had been recruited into the Soviet security-police network and assigned to Comintern responsibilities. A true "little Stalin," Chervenkov quickly aggregated in his own hands the plural powers of governmental prime minister, mass organizational chairman (Fatherland Front), and party secretary-general, simultaneously demoting and purging the wartime "local underground" Communists in favor of his fellow "Muscovites." He also inflicted on Bulgaria the full repertoire of mature Stalinist institutions and policies: terror, nationalization, industrialization, collectivization, cultural socialist realism, Russomania, and "cult of personality."

After the death of Stalin, the Soviet Union entered the phase of "collective leadership," whose imitation was at that time obligatory throughout the bloc. Hence Chervenkov had to divest himself of part of his plurality of offices, yielding the party leadership to the little known and widely underrated Todor Zhivkov in March 1954. He had also to tolerate the rehabilitation of other wartime "local underground" Communists who had been eclipsed since the anti-Kostov purge. And he further emulated the Malenkovist Soviet model by endorsing a shift of investment priorities toward agriculture, housing, consumption, and so-called cultural and social allocations in order to raise living standards. Indeed, Chervenkov sought to adapt not only to the collegial format of the new Soviet ruling style, but also to its tentative legalism and populism. Thus he authorized the release of many political prisoners, the closing of some internment and forced-labor camps, the redress of citizens' grievances, the greater accessibility of officials and rulers to the public, and the buying out of the Soviet shares in the notoriously exploitative joint companies (see Chapter 3, section 6). But alas for him, Chervenkov misjudged the power balance within the Soviet leadership, selecting Malenkov as his patron. Hence his relative domestic popularity availed him nothing when Khrushchev prevailed in Moscow and proceeded to court Tito, whose grudge against Chervenkov and Rákosi (the less flexible Hungarian "little Stalin") for their conduct after 1948 was utterly implacable.

On April 17, 1956, two months after Khrushchev's pivotal denunciation of Stalin at the Twentieth Congress of the Soviet Communist party, Chervenkov was replaced as Bulgaria' prime minister by his long-standing archrival, the "local underground" Communist Anton Yugov, whom he had humiliated but not eliminated during his own ascent over the corpses of the Kostovites in 1949 and 1950. Nevertheless, Chervenkov remained a member of the party's Politburo and became a deputy premier. Indeed, the widespread perception that he was now the victim of Yugoslav vengeance and Soviet intrigue earned him some popular sympathy and buttressed his domestic standing, while his successor

Yugov was remembered with bitterness as the baleful interior minister from 1944 to 1949, when the non-Communist political parties had been eviscerated and Nikola Petkov judicially murdered (see Chapter 3, section 7). Yet the rivalry between Chervenkov and Yugov, though intense, did not reach the pitch of the feud between Rákosi and Nagy in Hungary; the party was not split so profoundly or so publicly as to invite general turmoil or mass revolt.

Bulgaria now entered a period of relative thaw, permissiveness, and openness—though not on the scale then sweeping Poland and Hungary, and always under cautious and alert party control. The autonomy of the police was reined and that of the judiciary was pledged. Economic plans were shifted toward higher investments and intensification in agriculture and the light and food industries, as Bulgaria—in contrast to Romania—eventually endorsed the Soviet recommendation of an "international socialist division of labor" within CMEA. It became a major exporter of fresh and processed fruits and vegetables to the Soviet Union, Czechoslovakia, and East Germany, in return for energy and machinery imports. It also achieved a sharp shift of labor from "raw" agriculture into processed agriculture, agricultural industry, and other light industry.

Meanwhile, the seemingly drab and plodding first secretary, Zhivkov, was steadily consolidating and extending his patronage power behind the backs, as it were, of the flashier rivals Chervenkov and Yugov. By 1961, he was ready to make his move. Khrushchev's resumption of the attack on Stalin's methods and heritage at the Twenty-second Congress of the Soviet Communist party in October 1961 was Zhivkov's license to recall Chervenkov's earlier Stalinist past and to oust him on November 29, 1961, from the deputy premiership as well as from the party's Politburo and Central Committee—indeed, to eject him altogether from Bulgarian political life for his ostensible "mistakes and vicious methods," which had generated an "improper and artificial personality cult." A year later, on November 5, 1962, it was Yugov's turn to be dismissed from the premiership and expelled from the Central Committee for "crude violations of socialist legality" and "undermining the unity of the party" as well as for alleged personal dishonesty, cowardice, cruelty, rudeness, vanity, and incompetence. (This last accusation, at any rate, was untrue.) Many prisoners were amnestied, and Kostov was finally vindicated and exonerated, thirteen years after his destruction in the Stalinist anti-Tito hysteria of 1949 (see Chapter 4, section 3). This double disgracing of two men who, for better or worse, had been their country's leading political personalities for a dozen years exemplifies that compounding of ideological issues, political questions, power struggles, and personal vendettas that is characteristic of a political system to which the concept of legitimate opposition is alien.[14]

Henceforth Zhivkov combined in his own person the governmental and party leaderships. He promptly purged and subjected to his own control the security-police apparatus of the Interior Ministry, which, as in other Communist states, had been prone to behave as an intrusive, unsupervised, and unaccountable power in the land. And the exposure of the "Maoist" military-political conspiracy in April 1965 enabled Zhivkov to do the same to the armed forces. Like several of his counterparts, Zhivkov became de facto ruler for life. Eschewing the ostentatious megalomania and dandyism of Ceauşescu and Tito, he resembled more the modest and stubborn Kádár, though without his principled commitment to structural reform. Zhivkov's style of rule, autocratic yet accessible and with a common touch, reflected vividly the fact that Bulgaria is a country without a democratic political tradition but with an emphatically egalitarian social tradition, hostile to hierarchies and committed to universal education.

By May 1971, Zhivkov felt sufficiently confident and assured to replace the so-called Dimitrov Constitution of December 1947 with a new one, which elevated the country's ideological status to that of a "socialist state," explicitly identified the Communist party as its governing authority, and cited friendship with the Soviet Union as its guiding principle. Zhivkov now became formal head of state. In 1979, eleven years after Kádár, he introduced a paler and more cautious imitation of the Hungarian NEM (see section 2), favoring the hitherto suspect private peasant plots and welcoming hitherto scorned Western investments. (The private plots, though amounting to only one-eighth of the country's arable acreage, produced between one-quarter and one-half of its vegetables, meat, eggs, and dairy products.) Though the technological proficiency of Bulgaria's industry remained unimpressive and most of its products were uncompetitive on Western markets despite the watered-down NEM, the country's symbiotic integration into CMEA and the complementarity of its economy with that of the Soviet Union enabled Bulgaria to maintain the best long-term economic growth rate of all CMEA partners.[15] In effect, the relatively immature Bulgarian economy was buying today's security at the cost of tomorrow's likely obsolescence—an economically myopic but politically convenient trade-off.

Zhivkov's optimism and confidence also extended to the cultural arena, where he balanced the nauseating postwar Russophilism with a new exaltation of specifically Bulgarian patriotism. This entailed the reappreciation of the nation's past, its saints, heroes, and holidays, as well as the rehabilitation of some disdained sociopolitical forces, such as the medieval Orthodox church and the prewar agrarian movement. This official neonationalism appears to have resonated to authentic public opinion, which had become weary of the excessive adulation of all things Soviet and Russian. Though not as traditionally suspicious of Russia as some other peoples of East Central Europe, the Bulgarians craved recognition of their own worth and creativity. Of late, their youth, like that of neighboring countries, had become fascinated by all things Western. Nevertheless, Zhivkov's sponsoring of national pride and assertiveness did not imply any political alienation from the Soviet Union, which continued to be valued for the economic and military security that it provided to this small Balkan state, which still perceived itself as encircled by cool and potentially predatory neighbors: Turkey, Greece, and Yugoslavia. And for the Soviet Union, in turn, the value of Bulgaria's loyalty was much enhanced by the defections of Yugoslavia and Albania and the ongoing obstreperousness of Romania in the context of the Sino-Soviet rift. That is presumably why Moscow occasionally authorized the reassertion of another traditional expression of Bulgarian nationalism—the perennial aspiration to Macedonia, which had been a bone of contention between the Yugoslav and the Bulgarian Communist movements even during World War II (see Chapter 2, sections 5 and 7). These recurring flare-ups of rhetorical and academic skirmishes over Bulgarian irredentist claims to Macedonia usually coincided with the periodic troughs in Soviet–Yugoslav relations.

Two black marks against the Zhivkov regime of the 1980s must be recorded here. The first was its unsavory and apparently extensive involvement in the international traffic in drugs and arms, which helped to defray Bulgaria's hard-currency debt. The second was the brutal drive, begun late in 1984, to extirpate the ethnic distinctiveness of the country's Muslim Turkish minority, numbering about one-tenth of the population, and compel the Turks' assimilation into Bulgarian culture. Smaller minorities of Albanians, Armenians, and Gypsies were subjected to similar coercion, whose aim was to render Bulgaria a single-nationality country.

It is meet to close this section on a human note. In the 1970s, the widowed Zhivkov made a stab at nepotism. His talented, serious, and industrious daughter Lyudmila (born in 1942), who had studied for a year at Oxford University, was elevated to the party's Central Committee and then to its Politburo, with special responsibility for culture, science, and education. She spurred and may even have conceived the regime's new nationalism. But her death in 1981 cut short her incipient grooming to succeed her father, whose transcendent political value and achievement is stability.

5

Stalin's perception of the Soviet stake in East Central Europe at the close of World War II was delineated earlier (see Chapter 2, section 9; Chapter 3, section 1). What is the presumptive perception of his heirs four decades later? The starting point for productive speculation about this question is the recognition that the monolithic Communist bloc of 1947 is no more. It has been replaced by an agglomeration of Communist states, of which the Soviet Union remains the most powerful, prestigious, and paramount one, whose rulers share a verbal and probably a conceptual commitment to Marxism-Leninism while perceiving themselves as national leaders obliged to pursue national interests, but without being democratically accountable to their several national societies. This self-image as national leaders occasionally prompts the East Central European Communist rulers toward contentious behavior that would have been deemed intolerable in Stalin's years, such as economic beggar-thy-neighbor policies toward one another's countries, protesting one another's treatment of ethnic minorities, rejecting one another's diplomatic notes, raising the specter of territorial irredentism, and so forth. And they reconcile such nationalistic behavior with their professions of international ideological Leninism in a casuistic fashion worthy of Stalin himself. Just as he, in the 1920s and 1930s, brilliantly and brutally resolved the contradiction between the imperatives of world revolution and of Soviet state interests by defining the latter as an integral and crowning part of the international revolutionary ideology itself (via the formulas of "socialism in one country" and "the fatherland of all toilers"), so today's Communist rulers of the East Central European states seek to resolve their dilemmas of ideological legitimacy by defining their particular "national forms" as inherently "socialist in content." Pioneered most blatantly by Romania's Ceauşescu, this rationalization has been emulated by the other rulers.

Their several East Central European societies, in turn, have become more complex and technologically sophisticated since Stalin's time—a process that has generated pressures for decentralization, for nonideological pragmatism, for emancipation from repressive political constraints, and for contact with the freer, more productive, and technologically superior non-Communist half of Europe. These pressures are operative within and among these states as well as between them and the still hegemonic Soviet Union. This network of pressures and interests—sometimes common and sometimes conflicting—has undermined the deceptively sharp line that spuriously partitioned Europe in the late Stalinist era.

When that partitioning line between Communist and non-Communist Europe was imposed after World War II, the structural flaws and historical nemeses that were built into the Stalinist system were not yet perceived correctly by most Western scholars, who tended to be mesmerized by the imagery and the model of "political totalitarianism." In retrospect, this former academic fascination with supposedly monolithic totalitarianism is understandable, though no longer fashionable. After all, Stalin's readiness and ability to apply his and the Soviet Union's power on East Central Europe in order to impose near uniformity on the area's polities and societies did render plausible such a monocausal across-the-board generalization about the people's democracies as a single class of political institutions and processes. Only after Stalin and after his partial repudiation, and as political dilemmas, hesitations, and conflicts occasionally gripped the Soviet center itself, could the East Central European periphery's traditional diversities, historical particularities, and sovereign orientations reassert themselves. Indeed, in the case of Titoist Yugoslavia, this had occurred during Stalin's lifetime, in the late 1940s. Since then, the supposedly monolithic totalitarian character of the Soviet bloc has been subjected to serious challenge during every decade: from Poland and Hungary in the 1950s; from Romania, Czechoslovakia, and Albania in the 1960s; from Poland in the 1970s and the 1980s; and perennially from Yugoslavia.

These political challenges, which have been facilitated since the early 1960s by the Sino-Soviet rift, have entailed structural and institutional consequences, as governmental and political arrangements in the states of East Central Europe came to diverge from one another as well as from those of the Soviet Union. Thus by the mid-1980s, Poland was ruled by its military apparatus, Czechoslovakia by a dogmatic and Hungary by a pragmatic party gerontocracy (until the accession of Grósz's team in May 1988), Romania by a clan, Bulgaria and Albania by conventional party regimes, and Yugoslavia by a confederal arrangement. Moscow's policies (though not its goals) toward the area have also become quite differentiated. Hence the political study of East Central Europe is today more complicated than it was in the seemingly simple era of totalitarian Stalinism.

Every Soviet leader since Stalin has sought to find a viable formula for achieving stability within East Central Europe and cohesion between this area and the Soviet Union. And all have failed. Khrushchev's de-Stalinization provoked the strains that led to the Polish crisis and the Hungarian Revolution of 1956, from which he tried to recoil by revitalizing CMEA, only to lose Romania as a satellite. His attempted reconciliation with Yugoslavia and rift with China then prompted and facilitated the defection of Albania. Brezhnev's toying with détente, technology transfers from West to East, and credit-financed international trade led to the "Czechoslovak spring" of 1968 and to the chronic Polish instability of the 1970s, culminating in the Solidarity upheavals of 1980 and 1981. Yuri A. Andropov and Konstantin U. Chernenko continued Brezhnev's prohibition of deep structural reforms, in whose absence the various policy tinkerings mentioned earlier went for naught, for they still left the Communist states unable to capitalize optimally on the freshly imported Western technology. The overall result of all these half-measures and systemic rigidities has been to transform both the Soviet Union and East Central Europe into a single "Greater East European Co-Stagnation Sphere."[16] And CMEA has become a semibarter network for second-rate goods and raw materials, while quality products are reserved for exchanges with the West. Indeed, Gorbachev now finds his East Central European clients in even bleaker political and economic shape than the Soviet Union itself. The legitimacy levels of their regimes are lower and the performance of most of their economies poorer, and hence burdensome to the Soviet economy. And the upshot of their stagnant living standards, glaring social inequities, hideous environmental pollution, and continuing, Soviet-mandated political and structural inflexibility has been to feed a burgeoning, barely repressed, anti-Soviet and anti-Communist nationalism that has already compromised the military utility of Poland and may soon jeopardize that of the others as well. Indeed, the WTO—never a true alliance—is today perceived by Soviet leaders as even less of a real military asset than when it was founded in 1955.[17] And the very fact of the Soviet imperial hegemony over the area in an age when other empires have withered away has become an embarrassment to the Soviet Union vis-à-vis the West, the Third World, and even the international Communist movement.

The immediate Soviet priority in East Central Europe is political stability. But that stability depends on improved economic efficacy, which, in turn, requires the kinds of structural reforms, Western credits, and superior Western capital equipment that have shown themselves to be politically and psychologically destabilizing in Czechoslovakia and Poland. Stability or viability, control or reform? The dilemma is stark; the circle is vicious; the stalemate is palpable. Thus throughout the Soviet Union's East Central European empire, we witness the exquisite irony of a classic Marxian contradiction between a seething socioeconomic substructure and an ossifying political superstructure. It is difficult to exaggerate the potential critical nature of this contradiction, for Communist systems have always anchored their political legitimacy much more explicitly and lineally in their economic performance than have non-Communist ones. Hence economic miasma tends to have a direct political ripple effect.

Gorbachev's input toward a resolution or even a recognition of this contradiction has so far been quite unhelpful. The widespread expectation at his accession to the Soviet leadership in March 1985, that he would inject a healthy dose of imaginative and dynamic innovation into Soviet–East Central European relations, has now dampened considerably. While he does, indeed, place greater emphasis than did several of his predecessors on efficiency and rationality, he retains their traditional requirement of tight bloc coordination and discipline under Moscow's overall direction. His emphatic insistence on enhanced CMEA integration, which is not market-based, will surely limit the opportunities for the East Central European economies to free themselves from the debilitating Communist tradition of centrally planned and administered economic parameters. And planned economies will prove more difficult to integrate than would market-based economies, especially in the prevailing atmosphere of local nationalism. Though it must be obvious in the Kremlin that the price structures of intra-CMEA trade are irrational and that this irrationality hobbles growth rates and spreads malaise in several member states' economies, Gorbachev has so far chosen to respond to this sorry spectacle with nothing better than the same tough but inadequate formula used by his predecessors—discipline before (and probably in lieu of) flexible markets—a formula that at best only palliates and probably really aggravates the deep structural contradictions in these nations. Whatever the extent of Gorbachev's boat rocking within the

Soviet Union, his behavior toward the external clients has been quite conservative.

These systemic and policy-aggravated sources of political infirmity within East Central Europe and of strain between the area and the Soviet Union are further exacerbated by a series of impending leadership successions in the smaller states. Such successions, being constitutionally and juridically unregulated, tend to be unsettling, both before and after they actually occur. Czechoslovakia's Husák, Hungary's Kádár, Bulgaria's Zhivkov, and East Germany's Honecker are well over seventy years of age. Romania's Ceauşescu (born in 1918) is in poor health. And even though Poland's Jaruzelski, Hungary's Grósz, Czechoslovakia's Jakeš, and Albania's Alia do not yet confront these actuarial and medical imperatives, they too face the strong possibility of political turmoil catalyzed by a lethal fusion of personal ambitions and dismal economics. The Yugoslav collective leadership has still not managed to restore the political mastery that lapsed with Tito's death on May 4, 1980; that country's always fragile interethnic balance is more precarious today than at any time since the end of World War II. And with the possible exception of Kádár, all the East Central European leaders have so far shown political skill only in avoiding and postponing hard choices among competing priorities, not in making them (for example, economic reform versus political stability)—a pattern of procrastination that until now has been abetted by successive Soviet leaders as well. But Gorbachev's emphatic recent signals to the effect that the Soviet Union no longer can or will economically subsidize political stability in its junior partners, while insisting on the maintenance of this very stability, suggest that the luxury of choosing not to choose is running out on the East Central European leaders.

Despite all these problems, liabilities, and handicaps that the Soviet Union incurs from its continuing imposition of Communism on East Central Europe, there is no signal that Stalin's heirs are prepared to retreat from it, nor any flagging of their political will to dominate the area. The domestic Soviet imperatives that were operative at the close of World War II still apply: control of East Central Europe is a source of immense pride and security to the Soviets, the most tangible evidence of their great victory in that war; and that victory remains the most powerful legitimating experience of the Soviet Communist system, the most authentic bond between the Soviet elite and its Slavic peoples as well as within that elite. Hegemony over East Central Europe compensates the Russian people for their enormous sacrifices in the war and their enduring grievances since its end. It validates the Soviet system to itself. Any

Soviet yielding of the area not only would undermine the ideological claims of Communism to be the unfolding expression of an inexorable historical process and degrade the Soviet Union's credentials as a confident global power, but also would gravely jeopardize a basic internal Soviet consensus and erode the domestic security of the system itself (see Chapter 2, section 9).[18]

But herein lies another fateful contradiction and moral–political stalemate. The domestic Soviet imperatives that require the continued imposition of controlled Communist regimes on East Central Europe are precisely the ones that abort the local legitimacy of those regimes in their countries. Except in Yugoslavia and Albania, none of these regimes came to power through their own struggles and on the strength of their own resources. They were imposed and are still maintained by Soviet power, a condition that severely compromises their legitimacy among their several publics. These publics remain nationalistic, often religious, as well as self-consciously "European" and culturally "Western" in the sense of anti-Soviet—the youths who grew up under these regimes no less than their elders. Communism, being a function of Soviet hegemony, remains suspect as an alien creed and as a mere cult of power. This situation, in turn, feeds a revival of interest in the most recent and hence easily evocable period of independence, sovereignty, and dignity—the two interwar decades. For all their defects, vicissitudes, and shameful episodes, they tend to be recalled with increasing nostalgia as a time of true political legitimacy, with functioning legislatures, lively parliamentary life, a critical press, and a legal order not naked to arbitrary political invasions. What does this revived interest in the interwar era reflect and portend?

6

If we shift focus from the preceding macropolitical tour to a micropolitical look at the nitty-gritty dimension of contemporary politics in East Central Europe, we find ourselves making a quite different assessment of the classic problem of "continuity versus change" than was made by scholars in the 1950s. To our predecessors and teachers, the Stalinist imposition of monopolistic Communist rule appeared—quite understandably—to be a profoundly revolutionary rupture with earlier patterns, traditions, and histories. Today, we are more impressed with the survival and resurgence of political continuities from the interwar period in such dimensions as the styles and degrees of political participation, the operational codes and cultures of political elites, the processes of recruiting these political elites, their definitions of economic priorities, and so forth. This emphatic difference between the perspectives of the 1950s and those of the 1980s need not be surprising. The tension between the principles of continuity and of change is the motor of history, and great revolutions—for example, the Christianization of the Roman Empire, the French Revolution, and the Bolshevik Revolution—embody this tension in a particularly acute form. And the longer the time elapsed since the zenith of the revolutionary paroxysm, the more palpable become the threads of continuity, which not only survive but resiliently reassert themselves even after the seemingly most disruptive of revolutionary upheavals. Indeed, the philosophically most self-conscious historians of these great revolutions—Theodor Mommsen, Alexis de Tocqueville, E. H. Carr—have demonstrated their intellectual power most vividly in showing and explaining these dynamics. So, too, with East Central Europe, whose political patterns in the 1980s look more continuous with those of the 1930s than seemed conceivable in the midst of the revolutionary decades of the 1940s and 1950s.

Mass political participation today is reminiscent of the style of politics by plebiscite of the 1930s. Voters turn out in great numbers for *soi-disant* elections whose outcomes are known in advance. And on public holidays (whose dates are, admittedly, not necessarily identical with those of the interwar era), the citizens again march in huge flag-bearing and banner-carrying demonstrations past the draped reviewing stands from which the leaders exhort them to diligence, vigilance, and obedience. The alleged spontaneity and authenticity of these rituals is, if anything, even more spurious today than it was in the era of Edward Rydz-Śmigły, Gyula Gömbös, Milan Stojadinović, King Carol, Tsar Boris III, and their counterparts.

The term *rituals* is used advisedly, for today's East Central European regimes have turned their backs on the Stalinist requirement of authentic subjective acceptance and endorsement of a revolutionary leadership's claims and values. Like the rulers of the 1930s, those of the 1980s are satisfied with merely ritualistic, symbolic, overt—even pretended—gestures of ratification. In a sense, these modes of mass political participation are really a kind of depoliticization of public life. The real decision-making process remains just as remote from these spurious mass celebrations, just as much a monopoly of a small, self-selected political elite as it was in the 1930s.

Who is the current political elite? Not the direct genealogical descendants of the interwar bureaucratic-intelligentsia-gentry elite. That elite was toppled, or at least swamped, by the upheavals of World War II and the postwar massive, rapid Stalinist injection into the corridors of power of veteran Communist cadres and levies of newly radicalized "red-diploma" workers and peasants. Yet the old, ousted elite took its "Hegelian revenge" ("mankind makes its history behind its own back") by transmitting many of its styles, traits, and values to these usurpers, who, in turn, have made sure that their own children, rather than another generation of upstart workers and peasants, shall inherit their positions and privileges. Thus today, the corridors of power are as blocked to authentic workers and peasants as they were during the interwar era, while a highly politicized party and state bureaucracy mimics the gentry-intelligentsia pretensions of the interwar bureaucracy, apes its smugness, and replicates its nepotism, reigns, and rules. The socially closed nature of this self-protecting and self-replicating elite is redoubled by the prevailing patterns of access to higher education. All the ideological rhetoric supposedly favoring the children of manual workers and peasants has indeed become mere rhetoric, as the proportion of university students from these "toiler" backgrounds has declined in recent years. Not surprisingly, the moral and psychological gap of the 1930s between this elite and these masses has also replicated itself in the 1980s.

As regards a major functional responsibility of this elite—the setting of economic priorities in order to achieve rapid modernization—its choice of strategy, though consciously and deliberately imitated from the Soviet model, also happens to be an unacknowledged continuation of the policy of the bureaucrat-politicians of the 1930s in most of the area's states: to give top priority to industry, spearheaded by heavy metallurgy, while squeezing the necessary capital investments out of agriculture, consumption, and foreign credits. Nor is this particularly surprising, since the repertoire of possible paths to belated yet rapid modernization is, after all, limited, and the decision makers of the 1930s were no more free-market capitalistic entrepreneurs than are the contemporary ones; rather, both sets are state-capitalistic bureaucrats. State capitalism and state direction of the economy were quite extensively developed in interwar East Central Europe, and in this dimension, the policies of the postwar Communist regimes have expressed a high degree of strategic economic continuity within a setting of ideological change (see Chapter 1, section 7).

In one related socioeconomic dimension, however, the Communists have, for better and worse, "achieved" a goal that had eluded their interwar predecessors—the destruction of the peasant way of life that had immemorially characterized the elemental appearance, rhythm, and mores of East Central European social life and that had been both its bane and its solace. This destruction of a way of life was not simply a matter of statistically truncating the percentage of the population engaged in agriculture, but of transforming the whole character of a society, including even the character of its residual agrarian sector, which has become "post-peasantist" in its values and outlook. And those masses physically uprooted from village to town have, alas, been transformed from sullen and alienated peasants into sullen and alienated proletarians.

Despite many such similarities among the institutions and processes of the countries of contemporary East Central Europe, one should not necessarily infer that a uniform, regional political culture has been created by and under Communist power or that one existed before Communist regimes were established. Indeed, the persistence and resilience of distinct and diverse political cultures within the matrix of common Communist institutions is quite striking and lends support to the often maligned, much abused, and admittedly imprecise notion of national character. This, too, is a thread of continuity with earlier history. Without going into depth or detail, one can assert that, at a minimum, the citizens of each East Central European nation perceive their particular state as having a moral and historical significance far beyond being a mere unit in a supposed "socialist fraternity" of states and peoples. An old Leninist-Stalinist adage has been reversed: culture in contemporary East Central Europe may (or may not) be "socialist in form," but it is very much "nationalist in content."

Finally, in the sociopsychological dimension of political life, there is one more, alas unedifying, continuity from the 1930s, which has been ruefully marked by the Polish historian Jerzy W. Borejsza: "Sensibility to nationalistic and racial ideas, submission and subordination to any authoritarian power, and ruthlessness in combatting adversaries are all a legacy of that epoch."[19] The Communist apparats have inherited, adopted, refined, and intensified a deplorable tradition of conducting domestic politics not as an exercise in compromise and consensus building among fellow citizens, but as a mode of warfare against enemies.

Notes

Chapter 1

1. See, for example, Milan Hodža, *Federation in Central Europe* (London: Jarrolds, 1942), passim; David Mitrany, *Marx Against the Peasant* (New York: Collier Books, 1961), passim; Ghiţa Ionescu and Ernest Gellner, eds., *Populism* (New York: Macmillan, 1969), chap. 4.

Chapter 2

1. Hajo Holborn, *The Political Collapse of Europe* (New York: Knopf, 1954), passim.

2. On July 17, 1941, immediately after the German conquest of the Soviet share of interwar Poland, Hitler added East Galicia to the *Generalgouvernement* (though its population was preponderantly Ukrainian), but assigned the Białystok district, despite its overwhelmingly Polish population, to the personal administrative responsibility of the governor of East Prussia. Białystok had been Prussian between 1795 and 1807.

3. U.S. Congress, House Select Committee on Communist Aggression, *Communist Takeover and Occupation of Poland*, 83rd Cong., 2d sess., 1954, H. Rept. 2684, pt. 3, p. 13.

4. J. K. Zawodny, *Death in the Forest: The Story of the Katyn Forest Massacre* (Notre Dame, Ind.: University of Notre Dame Press, 1962), passim.

5. Vojtech Mastny, *The Czechs Under Nazi Rule* (New York: Columbia University Press, 1971), chaps. 4, 10.

6. Quoted in Anna Josko, "The Slovak Resistance Movement," in *A History of the Czechoslovak Republic, 1918–1948*, ed. Victor S. Mamatey and Radomír Luža (Princeton, N.J.: Princeton University Press, 1973), p. 372.

7. Vojtech Mastny, "The Beneš-Stalin-Molotov Conversations in December 1943: New Documents," *Jahrbücher für Geschichte Osteuropas* 20 (1972): 367–402.

8. Edward Taborsky, "The Triumph and Disaster of Eduard Beneš," *Foreign Affairs* 36, no. 4 (July 1958): 678.

9. Between September 1943 and March 1944, the royal government-in-exile located itself in Cairo, but then returned to London.

10. Martin van Creveld, *Hitler's Strategy, 1940–1941: The Balkan Clue* (London: Cambridge University Press, 1973), p. 176. In the summer of 1943, on the eve of Italy's surrender, 308,000 Italian troops were on occupation duty in various parts of Yugoslavia (Walter R. Roberts, *Tito, Mihailović, and the Allies, 1941–1945* [New Brunswick, N.J.: Rutgers University Press, 1973], p. 145).

11. Vladimir Dedijer, *Josip Broz Tito: Prilozi za Biografiju*, p. 260, quoted in Fitzroy Maclean, *Tito* (New York: Ballantine Books, 1957), p. 94.

12. The Ottoman Empire had ceded Bessarabia to Russia in 1812 and Bukovina to Austria in 1775. They were annexed to Romania after World War I. Southern Dobruja had been part of Bulgaria before the Second Balkan War of 1913, and Transylvania had been part of historic Hungary for a millennium until the end of World War I. As for the postwar outcome of Romania's surrender of all or parts of these four lands in 1940, the cession to Hungary was to be reversed after World War II but those to the Soviet Union and Bulgaria to be confirmed, thus negating Romania's attempted recovery between 1941 and 1944 of the lands it had yielded to the Soviet Union in 1940 and leaving it with a current post–World War II area of 237,502 square kilometers (91,676 square miles).

13. Adolf Hitler, *Hitlers Tischgespräche im Führerhauptquartier*, 3d, enlarged, ed. Henry Picker (Stuttgart: Seewald Verlag, 1976), pp. 178, 180, 209.

14. It has been suggested that the Western Big Two should have pressed their Yugoslav and Greek allies to concede these territories to Bulgaria, thereby enticing it into an early withdrawal from the war before the Soviet army reached the Balkans and thus rendering a "Finnish" rather than a Communist outcome plausible for it, which would have benefited Yugoslavia, Greece, and Turkey as well as Bulgaria itself (Nissan Oren, *Revolution Administered: Agrarianism and Communism in Bulgaria* [Baltimore: Johns Hopkins University Press, 1973], pp. 75–76). But this recommendation is sheer hindsight, overlooking the moral and psychological climate of wartime policy making and implicitly excusing the miserable Bulgarian behavior in the annexed lands.

15. King Zog spent the war years in England. But the British government (in contrast to its stance toward the Polish, Czechoslovak, Yugoslav, and other governments-in-exile) declined to extend to him diplomatic recognition or to accept any responsibility for facilitating his return home at war's close. After the war, Zog lived in Egypt and France, where he died on April 9, 1961.

16. The diary entry of the Bulgarian regent, Professor Bogdan Filov, for March 13, 1944, is pertinent here: "We have to remain loyal to Germany to the end. . . . The Italians not only did not gain anything but now are even held in contempt by the Anglo-Americans. The matter may come to an unconditional surrender, but it is necessary to be honorable" (quoted in Marshall L. Miller, *Bulgaria During the Second World War* [Stanford, Calif.: Stanford University Press, 1975], p. 171).

Chapter 3

1. Edward R. Stettinius, Jr., *Roosevelt and the Russians* (New York: Doubleday, 1949), pp. 314–15.

2. Alexander Dallin, "The Soviet Stake in Eastern Europe," *Annals of the American Academy of Political and Social Science* 317 (May 1958): 138–45; Zbigniew Brzezinski, *The Soviet Bloc*, rev. and enl. ed. (Cambridge, Mass.: Harvard University Press, 1971), pp. 4–9, 45–51.

3. Susanne S. Lotarski, "The Communist Takeover in Poland," in *The Anatomy of Communist Takeovers*, ed. Thomas T. Hammond (New Haven, Conn.: Yale University Press, 1975), p. 347.

4. In the course of this rapid expansion, Mikołajczyk's party was, alas, infiltrated by some unsavory people, which facilitated the Communists' campaign to delegitimate it. Indeed, some Communist-directed police provocation appears to have been involved. A case in point is the Kielce pogrom of July 4, 1946, in which forty-one Jews were murdered.

5. Pope Pius XII's stance was reversed by his successor, John XXIII, in March 1962.

6. In 1946, two more minor parties were permitted in Slovakia, but they did not significantly influence its political profile.

7. Though these elections were indeed free, one should note that voting was compulsory and that it could affect only the relative balances among the National Front parties; it was not possible to vote for any opposition party. Technically, the elections were for a constituent assembly that would both legislate conventionally and draft a new constitution.

8. The resignation of the Czechoslovak anti-Communist ministers was thus tactically an even more egregious blunder than was Mikołajczyk's questionable decision to draw the line on the issue of the Senate in the Polish referendum of June 1946.

9. Beneš withdrew to his country villa immediately after these events. He resigned the presidency three months later, on June 7, to avoid signing a Communist-drafted constitution, and died on September 3. He was succeeded in the presidency by the Communist leader Gottwald, who died on March 14,

1953—appropriately, from influenza caught while attending the funeral of his mentor, Stalin.

10. Josef Korbel, *The Communist Subversion of Czechoslovakia, 1938–1948* (Princeton, N.J.: Princeton University Press, 1959), pp. 212–13, 221–23; Jozef Lettrich, *History of Modern Slovakia* (New York: Praeger, 1955), pp. 236, 257–58; Dana Adams Schmidt, *Anatomy of a Satellite* (Boston: Little, Brown, 1952), pp. 109–10, 134–35; Jörg K. Hoensch, *Geschichte der Tschechoslowakischen Republik, 1918–1965* (Stuttgart: Kohlhammer, 1966), pp. 144–45.

11. Joseph Rothschild, *East Central Europe Between the Two World Wars* (Seattle: University of Washington Press, 1974), pp. 92–94, 135.

12. Like other ostensibly independent parties in neighboring states, these Czechoslovak "Potemkin" exhibits are now small, have a high ratio of functionaries to members, and are kept in existence only to sustain the fiction of a national front.

13. Archbishop Beran's internment lasted until February 1965, when he was permitted to leave for Rome on being designated a cardinal.

14. William O. McCagg, Jr., "Communism and Hungary, 1944–1946" (Ph.D. diss., Columbia University, 1965), chap. 4.

15. Ferenc Nagy, *The Struggle Behind the Iron Curtain* (New York: Macmillan, 1948), pp. 161–63.

16. After the Communist consolidation of power in 1948, the Soviet Union would reduce by half Hungary's outstanding reparations obligations.

17. These are rounded estimates rather than precise figures because the postwar Hungarian Central Statistical Office no longer publishes data on religious affiliation or preferences. But as there is no evidence of large-scale apostasy, they may be deemed proportionately correct.

18. For his own explanation, alleging physical abuse, psychological humiliation, and the administration of mind-altering drugs, see József Cardinal Mindszenty, *Memoirs* (New York: Macmillan, 1974), chaps. 4, 5.

19. Charles Gati, *Hungary and the Soviet Bloc* (Durham, N.C.: Duke University Press, 1986), pp. 5, 118, citing an unpublished section of the memoirs of Zoltán Vas, a leading "Muscovite" Hungarian Communist of those years; McCagg, "Communism and Hungary," p. 158.

20. Milovan Djilas, "The Tasks that Currently Confront Our Party," *Borba* (Belgrade), January 12, 13, and 14, 1946.

21. Proceeding from northwest to southeast, the six republics are Slovenia, Croatia, Bosnia-Hercegovina, Serbia, Montenegro, and Macedonia. Within Serbia, there are two autonomous provinces: Kosovo (originally Kosovo-Metohija) in the south, with an Albanian majority; and the Vojvodina in the north, with a Magyar, Romanian, Croatian, Serbian, Slovak, and Ukrainian population. Slovene, Serbo-Croatian, and Macedonian are designated as major languages, and Albanian, Turkish, Magyar, Slovak, Romanian, and Ukrainian as accepted ones in the two respective autonomous provinces.

22. In December 1951, after his rupture with Stalin and at a time when it was prudent to gratify the West, Tito released Stepinac to his native village. But a year later, he severed diplomatic relations with the Vatican in retaliation for Pope Pius XII's elevation of the archbishop to cardinal's rank—a gesture that Tito deemed a provocation. Stepinac died on February 10, 1960.

23. The only other civilian assigned a portfolio (Foreign Affairs) was a young nonpartisan career diplomat.

24. The able but unscrupulous new dissident Liberal minister of foreign affairs, Gheorghe Tătărescu, was so thoroughly tarnished by his long record of chronic opportunism, toying with anti-Semitism, dealing with the Iron Guard, and collaborating with Hitler that he was unlikely to make trouble for his Soviet protectors. In addition to the two dissident Liberals, the Groza cabinet included three renegade National-Peasants who were assigned less significant portfolios.

25. "The difference [between the Romanian "local" and "Muscovite" Communists] is somewhat difficult to define, since all were Leninists . . . and all believed in party discipline. Yet there was a difference, which may be illustrated by an example. Communists who had participated in the coup of August 23, were pleased that Rumania had switched sides so facilely and that the country had been spared the devastation of an invasion. Among the people coming from the East, however, there was some disappointment that the state structure had remained intact and they rather regretted that the issue had not been settled by force of arms, which would have cleared the way for immediate Communist reconstruction. This difference in view led to a number of arguments" (Henry L. Roberts, *Rumania: Political Problems of an Agrarian State* [New Haven, Conn.: Yale University Press, 1951], pp. 259–60). A charter member of the Communist party since its founding in 1921, a lawyer and social historian (that is, "a bourgeois intellectual"), and an ethnic Romanian, Pătrăşcanu also appears to have regarded the postwar autonomy extended to the Magyar minority as too generous. He was imprisoned from 1948 to 1954, then tried and executed in April 1954, to be posthumously rehabilitated in April 1968.

26. These so-called Sovroms were dissolved between 1954 and 1956, during the post-Stalin thaw, in belated Soviet acknowledgment of Romanian resentment at their colonial nature.

27. Quoted in Philip E. Mosely, *Face to Face with Russia*, Headline Series, no. 70 (New York: Foreign Policy Association, 1948), p. 23.

28. Joseph Rothschild, *The Communist Party of Bulgaria: Origins and Development, 1883–1936* (New York: Columbia University Press, 1959); Nissan Oren, *Bulgarian Communism: The Road to Power, 1934–1944* (New York: Columbia University Press, 1971).

29. The Radicals were admitted into the Fatherland Front shortly after the September 1944 *Putsch*; the Democrats were initially banned but then permitted to reorganize themselves in the autumn of 1945.

30. Phyllis Auty, "Bulgaria," in *Central and South East Europe, 1945–*

1948, ed. R. R. Betts (London: Royal Institute of International Affairs, 1950), p. 30.

31. On the relationship between Barnes and Petkov, see Michael M. Boll, *Cold War in the Balkans* (Lexington: University Press of Kentucky, 1984), chaps. 4–8.

32. Bruce Page, David Leitch, and Phillip Knightley, *The Philby Conspiracy* (New York: Doubleday, 1968), pp. 197–203; Patrick Seale and Maureen McConville, *Philby* (London: Hamish Hamilton, 1973), pp. 202–03.

33. Eric C. Browne, *Coalition Theories* (Beverley Hills, Calif.: Sage, 1973); Lawrence C. Dodd, *Coalitions in Parliamentary Government* (Princeton, N.J.: Princeton University Press, 1976).

34. Barrington Moore, Jr., *Soviet Politics: The Dilemma of Power* (Cambridge, Mass.: Harvard University Press, 1951), p. 375.

Chapter 4

1. In August 1946, the Yugoslavs forced down one, and then shot down another, American military transport plane over the northwestern corner of Yugoslavia. The planes were flying between Rome and Vienna. Tito initially responded with bristling truculence to American protests, until Molotov warned him to behave responsibly.

2. Charles Gati, *Hungary and the Soviet Bloc* (Durham, N.C.: Duke University Press, 1986), chap. 5; Ernst Halperin, *The Triumphant Heretic: Tito's Struggle Against Stalin* (London: Heinemann, 1958), p. 60.

3. Quoted in Nikita S. Khrushchev, "Cult of the Individual" (Secret speech delivered at the Twentieth Congress of the Communist party of the Soviet Union, February 25, 1956). In the event, the Tito regime found it necessary to arrest a mere 12,000 out of 470,000 party members for siding with Stalin and the Cominform.

4. Quoted in Zbigniew K. Brzezinski, *The Soviet Bloc: Unity and Conflict* (Cambridge, Mass.: Harvard University Press, 1971), p. 67.

5. Vladimir Dedijer, *Tito Speaks* (London: Weidenfeld and Nicolson, 1953), pp. 358, 360.

6. Adam B. Ulam, *Titoism and the Cominform* (Cambridge, Mass.: Harvard University Press, 1952), pp. 142–43.

7. The techniques of such "preparations," in which Soviet advisers participated, were later exposed during the "Czechoslovak spring" of 1968 by the few surviving defendants. An example is Eugen Löbl and Dušan Pokorný, *Hinter den Kulissen des Slánský Prozesses* (Vienna: Europa Verlag, 1968). Stalin's direct personal collusion in the Slánský purge was also documented in 1968 by the official Jan Piller Investigative Commission (H. Gordon Skilling, *Czechoslovakia's Interrupted Revolution* [Princeton, N.J.: Princeton University Press, 1976], pp. 397–404).

8. Pauker was not put on trial, but remained a private "nonperson" until her death in June 1960. Luca was tried on the usual charges in October 1954 (six months after the execution of Pătrăşcanu); his death sentence was commuted on appeal to life imprisonment, and he died in 1963. Bodnăraş remained on Gheorghiu-Dej's (and subsequently on Nicolae Ceauşescu's) leadership team, specializing in the political supervision of the armed forces and the security agencies, until his death in January 1976. Georgescu vanished for two decades, to be partially rehabilitated shortly before his death in the early 1970s.

9. The post-Stalin rehabilitation of Kostov began in April 1956 and was concluded with his full exoneration and vindication at the Eighth Communist Party Congress in November 1962.

10. Albania refrained from joining in this chorus of treaty denunciations in order to hold Tito to his formal obligation to protect it against Greek aspirations to northern Epirus (southern Albania). Tito put an end to this causistry by denouncing the Albanian-Yugoslav Treaty of Alliance on November 12, 1949.

Chapter 5

1. Andrzej Korbonski, *The Politics of Socialist Agriculture in Poland: 1945–1960* (New York: Columbia University Press, 1965), chaps. 8, 10.

2. Adam B. Ulam, *Titoism and the Cominform* (Cambridge, Mass.: Harvard University Press, 1952), p. 167.

3. Wyszyński had been freed by Gomułka on October 28. He remained the moral preceptor of the Polish nation until his death on May 28, 1981. Marshal Rokossovsky resumed his honored place in the Soviet army until his death on August 3, 1968. Many other Soviet officers seconded to the Polish army left with him in 1956.

4. The release by Nagy during the late summer of 1954 of some Communist political prisoners from Rákosi's post-1948 purges had an impact on the Hungarian party and police analogous to that of the simultaneous revelations of a senior Polish police defector.

5. At about this time, the Soviets also released the former executive secretary of the Smallholder party, Béla Kovács, whose arrest in February 1947 had signaled the final Communist drive to monopolize power (see Chapter 3, section 4).

6. Paul Kecskemeti, *The Unexpected Revolution: Social Forces in the Hungarian Uprising* (Stanford, Calif.: Stanford University Press, 1961), pp. 76–82.

7. Imre Nagy, *On Communism* (New York: Praeger, 1957), chaps. 1–3, 20–21; Zbigniew Brzezinski, *The Soviet Bloc: Unity and Conflict* (Cambridge, Mass.: Harvard University Press, 1971), pp. 219–22.

8. Luckily for Moscow, international attention was largely diverted from

its invasion of Hungary by the simultaneous Anglo-French attempt to seize Egypt's Suez Canal.

9. J. F. Brown, "Rumania Steps Out of Line," *Survey*, no. 49 (October 1963): 27. Between 1955 and 1964, the non-Communist world's share of Romania's foreign trade increased from 20 percent to 33 percent of the total, while the Communist world's share declined correspondingly from 80 percent to 67 percent. The Soviet Union's own particular share declined from 69 percent to 41 percent (George Gross, "Rumania: The Fruits of Autonomy," *Problems of Communism* 15, no. 1 [January–February 1966]: 23). By 1967, the Soviet Union's share was down to 27 percent (Paul Lendvai, *Eagles in Cobwebs* [New York: Doubleday, 1969], p. 338).

10. Ghiţa Ionescu, *Communism in Rumania, 1944–1962* (London: Oxford University Press, 1964), pp. 336–37.

11. Vladimir Tismaneanu, "Ceauşescu's Socialism," *Problems of Communism* 34, no. 1 (January–February 1985): 50, 62.

12. Galia Golan, *The Czechoslovak Reform Movement* (Cambridge: Cambridge University Press, 1971), pp. 4, 8. After his death on March 14, 1953, Klement Gottwald had been succeeded in the state presidency by Antonín Zápotocký, hitherto premier, and in the party first secretaryship by Antonín Novotný. After Zápotocký's death on November 13, 1957, Novotný held both offices.

13. The texts of these statements are in Robin A. Remington, ed., *Winter in Prague* (Cambridge, Mass.: MIT Press, 1969), Documents 29, 30, 34, 36, 38.

14. Ibid., Document 43.

15. In contrast to Poland's Gomułka and East Germany's Walter Ulbricht, Hungary's Kádár had been sympathetic toward the Czechoslovak experiment and participated only reluctantly in the invasion, which Romania's Ceauşescu and Albania's Hoxha denounced openly. In the autumn of 1956, incidentally, the shoe had been on the other foot. At that time, Novotný, anxious lest the Polish and Hungarian ferment spill over, had closed Czechoslovakia's borders with those two neighbors and jammed their radios. Another difference between 1956 and 1968 is that the Soviet military alone had invaded Hungary, whereas it was accompanied by units of four "fraternal allies" into Czechoslovakia (ostensibly a fifth "fraternal ally").

16. The text of the "Brezhnev Doctrine," which was actually articulatd not by Brezhnev but by a relatively minor Soviet ideologist in the pages of *Pravda*, September 26, 1968, is in Remington, *Winter in Prague*, Document 65.

17. Nicholas C. Pano, "The Albanian Cultural Revolution," *Problems of Communism* 23, no. 4 (July–August 1974): 50, 54.

18. Peter R. Prifti, "Albania's Expanding Horizons," *Problems of Communism* 21, no. 1 (January–February 1972): 34. Hoxha's behavior at this time made an interesting contrast to his endorsement, twelve years earlier, of the

Soviet suppression of the Hungarian Revolution, which he had interpreted as signaling an impending return to healthy Stalinism in the Soviet Union.

19. Elez Biberaj, *Albania and China: A Study of an Unequal Alliance* (Boulder, Colo.: Westview Press, 1986), pp. 145–49, 154–55; Nicholas C. Pano, "Albania: The Last Bastion of Stalinism," in *East Central Europe: Yesterday-Today-Tomorrow*, ed. Milorad M. Drachkovitch (Stanford, Calif.: Hoover Institution Press, 1982), p. 212. In October 1983, Albania and China resumed limited trade, exchanging Albanian chromium for Chinese spare parts for machinery and weapons.

20. Biberaj, *Albania and China*, pp. 100–03, 155–57.

21. Not all of Yugoslavia's Albanians reside in Kosovo, though that is where the political tension is most acute. Substantial and compact minorities live in Macedonia and Montenegro, and a scattered diaspora does unskilled, itinerant work in several cities throughout the country.

22. Dennison Rusinow, *The Yugoslav Experiment, 1948–1974* (Berkeley: University of California Press, 1977), pp. 91–94.

23. Brzezinski, *Soviet Bloc*, chap. 13; Stephen Clissold, ed., *Yugoslavia and the Soviet Union, 1939–1973: A Documentary Survey* (London: Oxford University Press, 1975), pp. 72–77.

24. A. Ross Johnson, "Yugoslavia and the Sino-Soviet Conflict: The Shifting Triangle, 1948–1974," *Studies in Comparative Communism* 7, nos. 1 and 2 (Spring–Summer 1974): 184–203.

25. Also illustrative of that balance were the conclusion of a commercial agreement with the European Economic Community on March 19, 1970, the restoration of full diplomatic relations with the Vatican in August 1970 (for their severance on the elevation of Alojzije Stepinac to the cardinalate in December 1952, see Chapter 3, section 5), and the first American presidential state visit to Yugoslavia, by Richard Nixon, on September 30 and October 1, 1970.

26. Rusinow, *Yugoslav Experiment*, p. 189.

27. Pedro Ramet, *Nationalism and Federalism in Yugoslavia, 1963–1983* (Bloomington: Indiana University Press, 1984), chaps. 5, 10.

28. Joseph Rothschild, *Ethnopolitics* (New York: Columbia University Press, 1981), chaps. 2–5.

Chapter 6

1. J. Michael Montias, "Economic Conditions and Political Instability in Communist Countries: Observations on Strikes, Riots, and Other Disturbances," *Studies in Comparative Communism* 13, no. 4 (Winter 1980): 286–89.

2. Other expressions of this "self-organization" process were KOR's broadened successor, the Committee for Social Self-Defense (KSS); the Movement

for the Defense of Human and Civil Rights (ROPCiO); the Society of Academic Courses (TKN), popularly dubbed "The Flying University"; the Independent Publishing Office (NOWa); the Polish League for Independence (PPN); the Confederation for an Independent Poland (KPN); the Young Poland Movement (RMP); the Student Solidarity Committees (SKS); and the Peasants' Self-Defense Organization. In the atmosphere generated by the Helsinki Final Act of 1975, with its emphasis on human rights, it would have been too embarrassing for the government to suppress such organizations.

3. Quoted in Nicholas G. Andrews, *Poland 1980–81: Solidarity Versus the Party* (Washington, D.C.: National Defense University Press, 1985), p. 197, and Peter Raina, *Poland 1981: Towards Social Renewal* (London: George Allen and Unwin, 1985), p. 321.

4. Jaruzelski's reticence to use terror continued a Polish Communist pattern noted earlier in connection with the Communist stances toward the Catholic church, the peasantry, and the party heretics in the late 1940s (see Chapter 3, section 2; Chapter 4, section 3).

5. Bill Lomax, "Hungary: The Quest for Legitimacy," in *Eastern Europe: Political Crisis and Legitimation*, ed. Paul G. Lewis (London: Croom Helm, 1984), p. 102.

6. Kádár's restraint in contrast to that of several of his peers in neighboring countries was also expressed by his refraining from formally designating his state as a "socialist republic" and retaining the ideologically more modest appellation of "people's republic."

7. Bennett Kovrig, "Hungarian Socialism: The Deceptive Hybrid," *Eastern European Politics and Societies* 1, no. 1 (Winter 1987): 118–24; Paul Marer, "Hungary's 1981–85 Five-Year Plan: Further Economic Reforms, Policy Changes, and Implications," in *War and Society in East Central Europe*, ed. Béla K. Király, Barbara Lotze, and Nándor F. Dreisziger (New York: Brooklyn College Press, 1984), vol. 11, pp. 487–98.

8. Zvi Gitelman, "The Politics of Socialist Restoration in Hungary and Czechoslovakia," *Comparative Politics* 13, no. 2 (January 1981): 187–210.

9. Michael Kraus, "Czechoslovakia and the Soviet Bloc: From Stalin to Gorbachev," in *The Political Economy of Bloc Cohesion*, ed. Cyril E. Black and Kenneth Oye. Forthcoming.

10. Milan Šimečka, *The Restoration of Order: The Normalization of Czechoslovakia* (London: Verso Editions, 1984), chap. 16; Václav Havel, "Letter to Dr. Gustáv Husák," *Survey* 21, no. 3 (Summer 1975): 167–90.

11. H. Gordon Skilling, "Independent Currents in Czechoslovakia," *Problems of Communism* 34, no. 1 (January–February 1985): 32–49.

12. In January 1948, Georgi Dimitrov had unintentionally provoked Stalin's ire by promoting the idea of a Balkan federation. But he had simply misread Stalin's priorities and was far from contesting him. Dimitrov promptly disowned his indiscretion.

13. Joseph Rothschild, *The Communist Party of Bulgaria: Origins and Development, 1883–1936* (New York: Columbia University Press, 1959), passim.

14. J. F. Brown, *Bulgaria Under Communist Rule* (New York: Praeger, 1970), chap. 6. Chervenkov, incidentally, now behaved with dignity, refusing to indulge in self-criticism. His party membership was quietly restored in 1969, and he died in obscurity in 1980, at age eighty.

15. John D. Bell, *The Bulgarian Communist Party from Blagoev to Zhivkov* (Stanford, Calif.: Hoover Institution Press, 1986), pp. 132–37, 146–47.

16. Seweryn Bialer, "The Genesis of Gorbachev's World," *Foreign Affairs* 64, no. 3 (1986): 629.

17. John Erickson, "The Warsaw Pact: From Here to Eternity?" *Current History* 84, no. 505 (November 1985): 359.

18. Seweryn Bialer, *The Soviet Paradox: External Expansion, Internal Decline* (New York: Knopf, 1986), pp. 191–93.

19. Jerzy W. Borejsza, "Italian Fascism and East-Central Europe: Problems of Typology and Links," in *Dictatorships in East-Central Europe, 1918–1939*, ed. Janusz Żarnowski (Wrocław: Ossolineum, 1983), p. 179.

Suggested Readings

Periodicals

Current History
East European Quarterly
Eastern European Politics and Societies
Problems of Communism
Studies in Comparative Communism
Survey

Books

General

Drachkovitch, Milorad M., ed. *East Central Europe: Yesterday-Today-Tomorrow*. Stanford, Calif.: Hoover Institution Press, 1982.

Fejtö, François. *A History of the People's Democracies*. New York: Praeger, 1971.

Gati, Charles, ed. *The International Politics of Eastern Europe*. New York: Praeger, 1976.

Griffith, William E., ed. *Communism in Europe*. 2 vols. Cambridge, Mass.: MIT Press, 1964, 1966.

Jelavich, Barbara. *History of the Balkans*. Vol. 2, *Twentieth Century*. Cambridge: Cambridge University Press, 1983.

Rakowska-Harmstone, Teresa, ed. *Communism in Eastern Europe*. 2d ed. Bloomington: Indiana University Press, 1984.

Terry, Sarah M., ed. *Soviet Policy in Eastern Europe*. New Haven, Conn.: Yale University Press, 1984.

Wolff, Robert Lee. *The Balkans in Our Time*. Cambridge, Mass.: Harvard University Press, 1956.

World War II

Amery, Julian. *Sons of the Eagle*. London: Macmillan, 1948.
Djilas, Milovan. *Wartime*. New York: Harcourt Brace Jovanovich, 1977.
Macartney, C. A. *A History of Hungary, 1929–1945*. 2 vols. New York: Praeger, 1956, 1957.
Maclean, Fitzroy. *Tito*. New York: Ballantine Books, 1957.
Mamatey, Victor S., and Radomír Luža, eds. *A History of the Czechoslovak Republic, 1918–1948*. Princeton, N.J.: Princeton University Press, 1973.
Mastny, Vojtech. *Russia's Road to the Cold War: Diplomacy, Warfare, and the Politics of Communism, 1941–1945*. New York: Columbia University Press, 1979.
Miller, Marshall L. *Bulgaria During the Second World War*. Stanford, Calif.: Stanford University Press, 1975.
Oren, Nissan. *Bulgarian Communism: The Road to Power, 1934–1944*. New York: Columbia University Press, 1971.
Roberts, Walter R. *Tito, Mihailović, and the Allies, 1941–1945*. New Brunswick, N.J.: Rutgers University Press, 1973.
Seton-Watson, Hugh. *The East European Revolution*. London: Methuen, 1956.
Zawodny, J. K. *Death in the Forest: The Story of the Katyn Forest Massacre*. Notre Dame, Ind.: University of Notre Dame Press, 1962.

The Communists Come to Power

Betts, R. R., ed. *Central and South East Europe*. London: Royal Institute of International Affairs, 1950.
Boll, Michael M. *Cold War in the Balkans*. Lexington: University Press of Kentucky, 1984.
Hammond, Thomas T., ed. *The Anatomy of Communist Takeovers*. New Haven, Conn.: Yale University Press, 1975.
Kertesz, Stephen D., ed. *The Fate of East Central Europe*. Notre Dame, Ind.: University of Notre Dame Press, 1976.
McCauley, Martin, ed. *Communist Power in Europe, 1944–1949*. New York: Barnes & Noble, 1977.
Markert, Werner, ed. *Jugoslawien*. Cologne: Böhlau, 1954.
———. *Polen*. Cologne: Böhlau, 1959.

Oren, Nissan. *Revolution Administered: Agrarianism and Communism in Bulgaria*. Baltimore: Johns Hopkins University Press, 1973.
Roberts, Henry L. *Rumania: Political Problems of an Agrarian State*. New Haven, Conn.: Yale University Press, 1951.
Schmidt, Dana A. *Anatomy of a Satellite*. Boston: Little, Brown, 1952.
Tönnes, Bernhard. *Sonderfall Albanien*. Munich: Oldenbourg, 1980.
Torańska, Teresa. *"Them": Stalin's Polish Puppets*. New York: Harper & Row, 1987.

The Dialectics of Stalinism and Titoism

Brzezinski, Zbigniew K. *The Soviet Bloc: Unity and Conflict*. Cambridge, Mass.: Harvard University Press, 1971.
Clissold, Stephen, ed. *Yugoslavia and the Soviet Union, 1939–1973: A Documentary Survey*. London: Oxford University Press, 1975.
Dedijer, Vladimir. *Tito Speaks*. London: Weidenfeld and Nicolson, 1953.
Halperin, Ernst. *The Triumphant Heretic: Tito's Struggle Against Stalin*. London: Heinemann, 1958.
Johnson, A. Ross. *The Transformation of Communist Ideology: The Yugoslav Case, 1945–1953*. Cambridge, Mass.: MIT Press, 1972.
Ulam, Adam B. *Titoism and the Cominform*. Cambridge, Mass.: Harvard University Press, 1952.

The Revenge of the Repressed

Bethell, Nicholas. *Gomulka: His Poland, His Communism*. New York: Holt, Rinehart and Winston, 1969.
Brown, J. F. *The New Eastern Europe: The Khrushchev Era and After*. New York: Praeger, 1966.
Floyd, David. *Rumania: Russia's Dissident Ally*. New York: Praeger, 1965.
Golan, Galia. *The Czechoslovak Reform Movement*. Cambridge: Cambridge University Press, 1971.
Griffith, William E. *Albania and the Sino-Soviet Rift*. Cambridge, Mass.: MIT Press, 1963.
Ionescu, Ghiţa. *Communism in Rumania, 1944–1962*. London: Oxford University Press, 1964.
Jowitt, Kenneth. *Revolutionary Breakthroughs and National Development: The Case of Romania, 1944–1965*. Berkeley: University of California Press, 1971.
Király, Béla, and Paul Jónás, eds. *The Hungarian Revolution of 1956 in Retrospect*. Boulder, Colo.: East European Monographs, 1978.

Korbonski, Andrzej. *The Politics of Socialist Agriculture in Poland, 1945–1960.* New York: Columbia University Press, 1965.
Lendvai, Paul. *Eagles in Cobwebs.* New York: Doubleday, 1969.
Lewis, Flora. *A Case History of Hope: The Story of Poland's Peaceful Revolutions.* New York: Doubleday, 1958.
Löwenthal, Richard. *World Communism: The Disintegration of a Secular Faith.* New York: Oxford University Press, 1966.
Montias, John Michael. *Economic Development in Communist Rumania.* Cambridge, Mass.: MIT Press, 1967.
Ramet, Pedro. *Nationalism and Federalism in Yugoslavia, 1963–1983.* Bloomington: Indiana University Press, 1984.
Rusinow, Dennison. *The Yugoslav Experiment, 1948–1974.* Berkeley: University of California Press, 1977.
Simmonds, George W., ed. *Nationalism in the USSR and Eastern Europe.* Detroit: University of Detroit Press, 1977.
Skilling, H. Gordon. *Czechoslovakia's Interrupted Revolution.* Princeton, N.J.: Princeton University Press, 1976.
Stehle, Hansjakob. *Eastern Politics of the Vatican, 1917–1979.* Athens: Ohio University Press, 1981.
Syrop, Konrad. *Spring in October: The Story of the Polish Revolution, 1956.* New York: Praeger, 1957.
Váli, Ferenc A. *Rift and Revolt in Hungary.* Cambridge, Mass.: Harvard University Press, 1961.
Weit, Erwin. *At the Red Summit: Interpreter Behind the Iron Curtain.* New York: Macmillan, 1973.
Windsor, Philip, and Adam Roberts. *Czechoslovakia 1968.* New York: Columbia University Press, 1969.
Zinner, Paul E. *Revolution in Hungary.* New York: Columbia University Press, 1962.

A Precarious Stalemate

Andrews, Nicholas G. *Poland 1980–81: Solidarity Versus the Party.* Washington, D.C.: National Defense University Press, 1985.
Ascherson, Neal. *The Polish August: The Self-Limiting Revolution.* New York: Penguin Books, 1981.
Ash, Timothy Garton. *The Polish Revolution: Solidarity.* New York: Scribner's, 1984.
Bell, John D. *The Bulgarian Communist Party from Blagoev to Zhivkov.* Stanford, Calif.: Hoover Institution Press, 1986.
Brown, J. F. *Bulgaria Under Communist Rule.* New York: Praeger, 1970.
Nelson, Daniel N., ed. *Romania in the 1980s.* Boulder, Colo.: Westview Press, 1981.

Raina, Peter. *Poland 1981: Towards Social Renewal*. London: George Allen and Unwin, 1985.

Šimečka, Milan. *The Restoration of Order: The Normalization of Czechoslovakia*. London: Verso Editions, 1984.

Stanković, Slobodan. *The End of the Tito Era: Yugoslavia's Dilemmas*. Stanford, Calif.: Hoover Institution Press, 1981.

Weschler, Lawrence. *The Passion of Poland: From Solidarity Through the State of War*. New York: Pantheon, 1982.

Surveys of Individual Countries

Gati, Charles. *Hungary and the Soviet Bloc*. Durham, N.C.: Duke University Press, 1986.

Grothusen, Klaus-Detlev, ed. *Jugoslawien*. Göttingen: Vandenhoeck & Ruprecht, 1975.

———. *Rumänien*. Göttingen: Vandenhoeck & Rupprecht, 1977.

———. *Ungarn*. Göttingen: Vandenhoeck & Rupprecht, 1987.

Hiscocks, Richard. *Poland: Bridge for the Abyss?* London: Oxford University Press, 1963.

Hoensch, Jörg K. *Geschichte der Tschechoslowakischen Republik, 1918–1965*. Stuttgart: Kohlhammer, 1966.

King, Robert R. *A History of the Romanian Communist Party*. Stanford, Calif.: Hoover Institution Press, 1980.

Kovrig, Bennett. *Communism in Hungary: From Kun to Kadar*. Stanford, Calif.: Hoover Institution Press, 1979.

Lange, Klaus. *Grundzüge der Albanischen Politik*. Munich: Trofenik, 1973.

Leslie, R. F., ed. *The History of Poland Since 1863*. Cambridge: Cambridge University Press, 1983.

Pano, Nicholas C. *The People's Republic of Albania*. Baltimore: Johns Hopkins University Press, 1968.

Prifti, Peter R. *Socialist Albania Since 1944*. Cambridge, Mass.: MIT Press, 1978.

Roos, Hans. *A History of Modern Poland*. New York: Knopf, 1966.

Shoup, Paul. *Communism and the Yugoslav National Question*. New York: Columbia University Press, 1968.

Stehle, Hansjakob. *The Independent Satellite: Society and Politics in Poland Since 1945*. New York: Praeger, 1965.

Suda, Zdeněk L. *Zealots and Rebels: A History of the Communist Party of Czechoslovakia*. Stanford, Calif.: Hoover Institution Press, 1980.

Weydenthal, Jan B. de. *The Communists of Poland: An Historical Outline*. Stanford, Calif.: Hoover Institution Press, 1978.

Index